West Chester Railroad

In April 1997, for the first train placed in service on the line since its closure in 1986, West Chester Railroad 65-ton diesel electric switcher locomotive No. 9 is at the West Chester, Pennsylvania, station platform. Built by the General Electric Company, this was actually an upgraded GE 44-ton locomotive with a heavier frame and more powerful diesel engine. On the right is former Reading Railroad multiple unit coach No. 9124, one of five coaches on the West Chester Railroad roster built by Harlan & Hollingsworth of Wilmington, Delaware, between 1931 and 1932. These coaches have had their traction motors and pantographs removed. (*Skip Small collection*)

Glen Mills, Pennsylvania, station is the location of West Chester Railroad locomotive No. 4230 on October 20, 2019. The station was built during 1880–1881 and became part of the Southeastern Pennsylvania Transportation Authority (SEPTA) West Chester line. SEPTA's last day of passenger service on the Elwyn-West Chester portion of the line was on September 18, 1986 due to deteriorating track conditions, low ridership, and Chester County's desire to expand facilities at the Exton station on SEPTA's Paoli-Thorndale line. (*Kenneth C. Springirth photograph*)

WEST CHESTER RAILROAD

KENNETH C. SPRINGIRTH

FONTHILL

On October 16, 2019, West Chester Railroad 2,400-horsepower type C424 diesel locomotives Nos. 4213 and 4230 are at the borough of West Chester in Chester County, Pennsylvania. Both of these locomotives were built by the Montreal locomotive Works in 1965 for the Canadian Pacific Railway. This railroad is a 100 percent volunteer operation. (*Kevin Knobl photograph*)

On top portion of cover: In June 1998, West Chester Railroad 1,750-horsepower type GP9 four-axle diesel electric locomotive No. 99 (built in October 1956 as No. 6499 for the Baltimore & Ohio Railroad by the Electro-Motive Division [EMD] of General Motors Corporation in the United States) is at the Queen Anne gothic architecture style Glen Mills, Pennsylvania, station (building on the right) of the West Chester Railroad. This type of locomotive was produced in the United States between January 1954 and August 1963. (*J. Nye photograph*)

On bottom portion of cover: West Chester Railroad locomotive No. 1803 is at the West Chester, Pennsylvania, yard waiting for the next assignment on October 19, 2019. The 1,800-horsepower type RS-18 diesel electric locomotive was built by Montreal Locomotive Works in April 1958 for the Canadian Pacific Railway and was acquired in 1998. On the right is locomotive No. 4230. (*Kenneth C. Springirth photograph*)

Back cover: On June 15, 2019, diesel locomotive No. 7706 is south of West Chester, Pennsylvania, on the historic former Pennsylvania Railroad West Chester via Media to Philadelphia rail line. The type GP38 diesel locomotive was built in 1969 by EMD for the Penn Central Transportation Company and was acquired in 2011. (*Andrew L. Keates photograph*)

Fonthill Media Inc.
www.fonthillmedia.com
office@fonthillmedia.com

First published 2021

ISBN 978-1-63499-306-7

Typeset in Utopia Std
Printed and bound in England

Contents

Acknowledgments

Beth Keates was outstanding for being on-site editor, contacting WCRR volunteers for pictures/information, and proofreading, making this book possible plus being a WCRR volunteer. This book is dedicated to Joseph Giacchino, owner of 4 States Railway Service that owns the railroad operating under the name West Chester Railroad (WCRR) plus helps maintain and operates locomotives; his wife, Ann Giacchino, who handles all of the special trips on the railroad; the four original founders (Skip Small who has handled track work, Don Callander, former director of the Wilmington & Western Railroad, Brian Woodcock past president of the Wilmington & Western Railroad, and Joseph Giacchino), former Congressman Curt Weldon, former Pennsylvania Senator Robert Thompson, former Pennsylvania State Representative Elinor Z. Taylor, former Pennsylvania State Representative Carol McElven-Smith, and all of the WCRR volunteers who do whatever is needed to keep the railroad in operation. The West Chester Railroad Heritage Association (WCRHA) made it possible for the Southeastern Pennsylvania Transportation Authority (SEPTA) to lease the railroad line to the WCRR.

Thanks to John Anshant, Glen Boekell, Dennis Boettcher, Charlie Brogan, Jimmy Butler, Chuck Denlinger, Al Draper, Jose Escamilla, Russ Farra, John Fernandes, Jen Fralick, Jen Fredrick, Ron Fricker, Steve Getz, Al Grover, Kristian Hamilton, Tyler Haney, Fred Heilich III, Rick Hess, Roger Holmes, Vince Jakubowski (trip coordinator for early SEPTA trips), Larry Jensen, John Jeppi, Guy Johnson, Andrew L. Keates, Jacob A. Keates, Beth Anne Keates, Brian A. Keates, Dave Kent, Kevin Knobl, Ashton Krug, Stephen Krug, Joe Lake, Kenny Levin, Jen Lipschultz, Clark Marshall, Melissa Magee (Channel 6 ABC News), Jimmy/J. C. McHugh, Louie Montanaro, Peggy Morrison, Frank Moscatelli, Jack Nolan, J. Nye, Joe Parsons, Johanna Rhoads, Steve Rogers, Glenn Schwass, Derek Slifer, Bryan Small, Judy Small, Skip Small, Dan Steil, Ron Stevens, Becky Stevens, Beth Stevens, Walter Snyder, Walt Still, Dave Swank, Ernie Taggart, Ed Thorton, Cecily Tynan (Channel 6 ABC News), Dave Walter, Heather Wiasuk (Channel 6 ABC News), Trip Williams, Samuel Wills, Joe Wilton, and Marc Zimmerman for all of their work on the West Chester Railroad. Thanks to Knox Equipment Rental and Summit Steel & Equipment Company. Thanks to Bob Elder of the Tunnel Inn at Gallitzin, Pennsylvania, for his hospitality and willingness to let us use his facility as this book was being planned and put together. Outstanding reference books were *Railroads of West Chester* by Jim Jones and *Riding the West Chester Railroad* by Jim Jones and Dave Walter plus various issues of *Newstracks* a publication of the West Chester Railroad Heritage Association. Schedules were used from the *June 1916, August 1936, and April 1952 Official Guide of the Railways*. Thanks to past and present WCRR volunteers who rebuilt the railroad between West Chester and Glen Mills.

In 2000, Mr. Joseph Giacchino is smiling with pride on the accomplishments of the West Chester Railroad Heritage Association, Four States Railroad Company, and all of the dedicated WCRR volunteers who have placed it back into operation. (*Skip Small collection*)

Introduction

In Pennsylvania, the West Chester Railroad completed a 9-mile line (using horse-drawn railcars) from West Chester with stations at West Chester (corner of Chestnut and Matlack Streets), Buchanan's Lane, Patton's Farm, Boot Road, Halfway House, White Horse Tavern, Ton Road, Hood's Road, and Intersection on September 13, 1832 where it intersected with the Philadelphia & Columbia Railroad. On November 7, 1832, the first train from West Chester left at 8 a.m. and arrived in Philadelphia at noon. A passenger could leave Philadelphia in the afternoon and get back that same day in West Chester by 8 p.m. In 1836, the West Chester station was relocated to Gay Street between High Street and Walnut Street. Steam locomotives were introduced in 1845.

With plans to build a railroad from Philadelphia via Media to West Chester, the West Chester & Philadelphia Railroad (WC&P) in August 1851 hired T. E. Sickels as the project engineer. The Malone, Clark & Gonder construction firm received the contract with a bid of $585,000 of which $200,000 was in company stock. This line represented engineering challenges requiring major bridges over Ridley, Crum, Darby, and Cobbs Creeks. Construction work began on April 10, 1852 at both ends of the line. At the end of 1852, there were 10.2 miles of graded right of way and 3.5 miles of track in place. However, the company missed a mortgage payment in April 1854 and went into receivership. The line began to haul freight and passengers from Philadelphia to Media in October 1854. Court appointed trustees Joseph T. Thomas and John Thomas took over on April 25, 1856. Between April 1856 and November 1858, they reorganized the company's books, consolidated the company's debt, and construction of the line to West Chester was completed. Engineer Edward Miller brought WC&P's first train from Philadelphia to West Chester on November 11, 1858 powered by Atlantic type steam locomotive Rockdale. Regular service began in December 1858. The Pennsylvania Railroad purchased a controlling interest in the Philadelphia, Wilmington & Baltimore Railroad in October 1881.

The Hoopes Brothers & Darlington Wheel Works (HBD) was located on Market Street next to the WC&P railroad station. HBD, the oldest manufacturing plant in West Chester, used the West Chester Railroad to receive carloads of lumber and to ship out wheels and wood products. As noted in the book *Railroads of West Chester* by Jim Jones, freight shipments sent from West Chester via the Media–West Chester line increased from 4,984 tons for the year ending October 31, 1889 to 8,029 tons for the year ending October 31, 1898. For that same time period, shipments received increased from 3,867 tons to 4,238 tons.

When the United States entered World War I and later World War II, railroads across the country handled large increases in passenger and freight loads. West Chester's downtown Market Street station handled a record amount of freight. Family members would come to the station to see their sons leave for military duty and were down to meet them on their return. After World War II, passenger train ridership declined. Beginning July 1953, passengers travelling between West Chester and Philadelphia had to change trains at Media. With the Philadelphia Suburban Transportation Company trolleys providing direct trolley service to 69th Street Terminal, trolley ridership increased. Initially, even with the conversion the West Chester trolley line to bus operation, ridership held fairly steady until a few years later when public transit ridership declined. On January 19, 1958, Sunday train service to West Chester was eliminated. As train fares increased and number of West Chester passenger trains declined, more riders began to drive.

Passenger train service continued to West Chester when the Pennsylvania Railroad and New York Central Railroad merged into the Penn Central Transportation Company in 1968 followed by bankruptcy and take over by Consolidated Rail Corporation in 1976. The Southeastern Pennsylvania Transportation Authority (SEPTA) began operating the Media–West Chester line in 1983. The last day of passenger train service to West Chester was on September 18, 1986 when the line was shortened to operate between Elwyn and Philadelphia. SEPTA noted that only a small number of riders were using the line and major track repairs were needed.

On June 19, 1996, 4 States Railway Service leased the West Chester to Glen Mills' portion of the line from SEPTA. The October 8, 1997 Federal Register page 52,616 noted 4 States Railway Service filed to sublease and operate 6.405 miles of rail line from the Borough of West Chester between milepost 27.4 and milepost 20.995 at Glen Mills station, Glen Mills, Delaware County, Pennsylvania. The 100 percent WCRR volunteer group with an amazing amount of hard work in a year reopened the line for passenger excursion service on September 21, 1997. They built a new station at 230 East Market Street in downtown West Chester as well as a new railroad yard for their acquired locomotives and cars. To the best of our knowledge, their 1927 built baggage car is the oldest continuously surviving operating Pennsylvania Railroad baggage car. The former Reading commuter passenger coaches later used by SEPTA are from the 1930s.

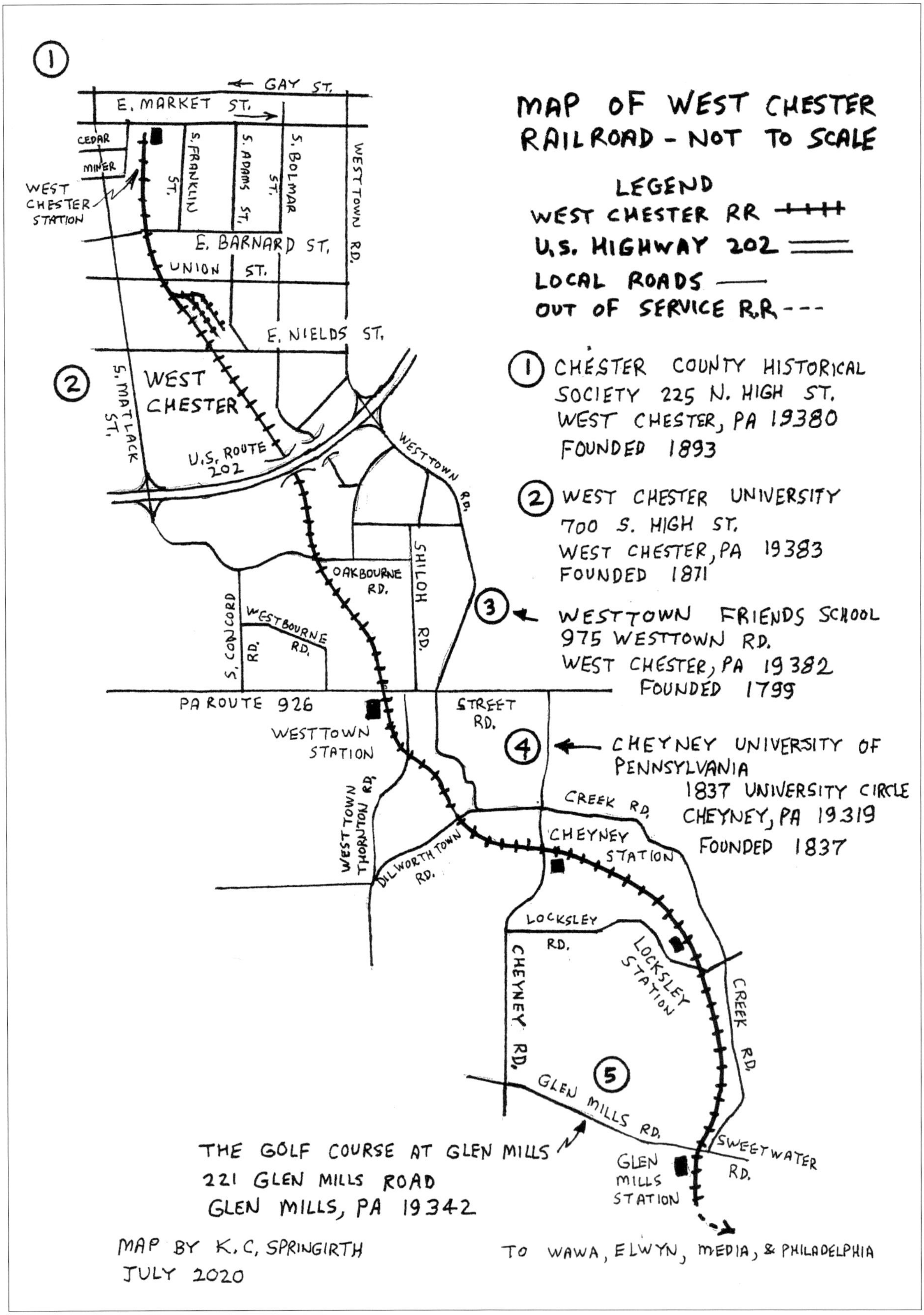

This is a map showing the West Chester Railroad operating between West Chester and Glen Mills, Pennsylvania.

1

Media–West Chester Line History

West Chester's first railroad, the 9-mile West Chester Railroad (WCRR), opened in September 1832 from West Chester to Intersection later known as Malvern where there was train service to Philadelphia via the Philadelphia & Columbia Railroad. In an era when people were anxious to have railroad competition, West Chester's second railroad the West Chester & Philadelphia Railroad (WC&P) began construction at both ends on April 10, 1852 connecting Philadelphia via Media to West Chester. By October 1854, passenger and freight service began between Philadelphia and Media. The WC&P's first passenger train from Philadelphia reached West Chester on November 11, 1858. In 1857, the Pennsylvania Railroad purchased the Philadelphia & Columbia Railroad and leased the WCRR for five years beginning April 1859 and ending March 1864. The WC&P set passenger fares and freight rates below the WCRR rates, which diverted business away from the WCRR. With the Philadelphia & Baltimore Central Railroad (P&BC) seeking a route into Philadelphia, it connected with the WC&P at Wawa. On April 1, 1864, the WC&P leased the WCRR, and track was laid east of Matlack Street from Biddle Street to Market Street where connection was made to the WC&P. In August 1879, the WC&P purchased the WCRR. The Pennsylvania Railroad (PRR) purchased the WCRR on August 1, 1879, and by 1881 acquired the WC&P. In 1887, the PRR brought a total of 238,287 passengers in and out of West Chester.

While West Chester has had six different passenger station locations, its best known station was located on East Market Street between Matlack and Franklin Streets. In 2020, this same location is used by the West Chester Railroad for its excursion trains. By 1857, the WC&P built a two-story brick train station at this location. The first train arrived at this station on November 11, 1858. As noted by *Railroads of West Chester*, written by Jim Jones, "A Sunday morning fire at the Market Street station on January 2, 1885 provided the pretext to remodel it once more. The company installed galvanized screens on the waiting room windows, built a shelter for the watchman who worked the crossing gates, and created a bunk room for visiting crews on the second floor."

Following successful electrification of the Paoli line on September 11, 1915, the Chestnut Hill West line was electrified on March 30, 1918, the West Chester line was electrified on December 2, 1928, and the Philadelphia to Wilmington line was electrified on September 30, 1928. Passenger service ended on the West Chester to Malvern line on September 24, 1932. While World War I and II resulted in increase in passengers and freight carried, after the war, ridership declined, and passenger service was reduced. Direct service between Philadelphia and West Chester ended in July 1953 with passengers changing trains at Media. Sunday train service from Media to West Chester ended on January 19, 1958. SEPTA's October 27, 1974 and April 27, 1980 schedules showed some trains were through trains.

As ridership declined, by 1962, only the ticket office remained in the station, and the walls had been defaced. When West Chester borough leaders asked the PRR to replace the station in 1965, the PRR applied for permission to demolish the station, which occurred in June 1968. The Pennsylvania Railroad and New York Central Railroad merged into the Penn Central Transportation Company on February 1, 1968, which later constructed a passenger shelter at Nields Avenue in West Chester and named it West Chester University station. That merger was not successful and Consolidated Rail Corporation (Conrail) took over Penn Central Transportation Company and six other railroads on April 1, 1976. Conrail provided commuter service under contract to the Southeastern Pennsylvania Transportation Authority (SEPTA) until January 1, 1983 when SEPTA took over operation of commuter service.

The last day of passenger service to West Chester was September 18, 1986, and the line was cut back to Elwyn. SEPTA had limited funds to make track repairs and offered Chester County the option of repairing the West Chester line or expanding parking at the Exton station. Chester County chose to expand parking at the Exton station. The West Chester Railroad has leased a 6.405-mile portion of the line between West Chester and Glen Mills from the Southeastern Pennsylvania Transportation Authority (SEPTA) since June 19, 1996. In 2018, SEPTA began construction work to restore service from Elwyn to Wawa.

437

PENNSYLVANIA SYSTEM

PENNSYLVANIA RAILROAD

Adams Express Company

Table 29—PHILADELPHIA DIVISION.

Local Trains Philadelphia to West Chester and Parkesburg.

May 28, 1916.	Mls	5303	5311	5501	5313	5315	5317	15	5319	5321	5323	5325	5329	5331
Philada....lve.	....	A M	A M	A M	A M	A M	A M	A M	▲	A M	A M	A M	A M	A M
Broad St...	0	†609	†625	†645	†715	†745	†815	†840	†845	*915	*945	*1015	*1045	†1115
West Phila. ⊡	1.1	6 13	6 29	6 49	7 19	7 49	8 19	8 44	8 49	9 19	9 49	10 19	10 49	11 19
Fortieth St.	2.7	- -	- -	- -	- -	- -	- -	- -	- -	- -	- -	- -	- -	- -
52d Street..	4.1	6 18	6 35	- -	7 24	7 54	8 24	- -	8 54	9 24	9 54	10 24	10 54	11 24
Overbrook..+	5.6	6 21	6 39	- -	7 27	7 57	8 27	- -	8 57	9 27	9 57	10 27	10 57	11 27
Merion.....+	6.1	6 23	6 41	- -	7 29	7 59	8 29	- -	8 59	9 29	9 59	10 29	10 59	11 29
Narberth...+	7.0	6 26	6 43	- -	7 32	8 02	8 32	- -	9 02	9 32	1002	10 32	11 02	11 32
Wynnewood+	7.6	6 28	6 45	- -	7 34	8 04	8 34	- -	9 04	9 34	1004	10 34	11 04	11 34
Ardmore...+	8.6	6 30	6 49	7 03	7 36	8 06	8 36	- -	9 06	9 36	1006	10 36	11 06	11 36
Haverford...+	9.3	6 32	6 51	- -	7 38	8 08	8 38	- -	9 08	9 38	1008	10 38	11 08	11 38
Bryn Mawr.+	10.3	6 35	6 57	7 08	7 41	8 11	8 41	- -	9 11	9 41	1011	10 41	11 11	11 41
Rosemont..+	11.0	6 37	6 59	- -	7 43	8 13	8 43	- -	9 13	9 43	1013	10 43	11 13	11 43
Villa Nova.+	12.1	6 39	7 02	- -	7 45	8 15	8 45	- -	9 15	9 45	1015	10 45	11 15	11 45
Radnor.....+	13.1	6 42	7 05	- -	7 48	8 18	8 48	- -	9 18	9 48	1018	10 48	11 18	11 48
St. David's.+	13.9	6 44	7 07	- -	7 50	8 20	8 50	- -	9 20	9 50	1020	10 50	11 20	11 50
Wayne.....+	14.6	6 46	7 09	7 16	7 52	8 22	8 52	- -	9 22	9 52	1022	10 52	11 22	11 52
Strafford....+	15.5	6 48	7 12	7 19	7 54	8 24	8 54	- -	9 24	9 54	1024	10 54	11 24	11 54
Devon......+	16.6	6 51	7 15	7 22	7 57	8 27	8 57	- -	9 27	9 57	1027	10 57	11 27	11 57
Berwyn....+	17.7	6 53	7 19	7 25	7 59	8 29	8 59	- -	9 29	9 59	1029	10 59	11 29	11 59
Daylesford...	18.7	6 55	7 22	7 28	8 01	8 31	9 01	- -	9 31	1001	1031	11 01	11 31	12 01
Paoli.......+	20.0	7 02	7 25	7 31	8 04	8 34	9 04	9 18	9 34	1004	1034	*1109	*1134	12 04
Green Tree.+	21.1	7 05	AM	7 34	AM	AM	AM	•925	AM	A M	A M	†1112	§1141	Noon
Malvern....+	21.7	7 07	....	7 36	....	....	....	•927	....	....	....	11 14	11 43	....
Frazer...arr.	24.0	7 12	....	7 42	....	....	....	9 25	....	....	....	11 19	11 48	....
Frazer+.lve.	24.0	7 12	....	8 10	....	....	....	9 32	....	....	....	11 19	11 48	....
Villa Maria. (W. O. Branch)	24.7	7 14	....	8 12	....	....	....	9 34	....	....	....	11 21	11 50	....
Morstein. +	25.7	7 17	....	8 15	....	....	....	9 37	....	....	....	11 24	11 53	....
Kirkland...	27.0	7 20	....	8 18	....	....	....	9 40	....	....	....	11 27	11 56	....
Green Hill..+	28.0	7 23	....	8 21	....	....	....	9 43	....	....	....	11 30	11 59	....
Fern Hill...	29.4	7 26	....	8 24	....	....	....	9 46	....	....	....	11 33	12 02	....
Maple Ave..	30.3	7 28	....	8 26	....	....	....	9 48	....	....	....	11 35	12 04	....
W. Chester+ar.	30.8	7 30	....	8 28	....	....	....	9 50	....	....	....	†1137	§1206	....
Glen Loch..+	25.5	A M	....	7 45	....	....	....	- -	....	....	....	A M	Noon	....
Ship Road....	26.8	....	....	7 48	....	....	....	- -	....	....	....	....	....	....
Whiteland....	27.5	....	....	7 51	....	....	....	- -	....	....	....	....	....	....
Whitford...+	28.5	....	....	7 54	....	....	....	- -	....	....	....	....	....	....
Bradford Hills	29.9	....	....	7 57	....	....	....	- -	....	....	....	....	....	....
Woodbine..+	31.2	....	....	8 00	....	....	....	- -	....	....	....	....	....	....
Downingtown.+	32.6	....	....	8 05	....	....	....	9 37	....	....	....	....	....	....
Thorndale..+	35.0	....	....	8 10	....	....	....	- -	....	....	....	....	....	....
Cain.........	36.8	....	....	8 13	....	....	....	- -	....	....	....	....	....	....
Coatesville.+	38.6	....	....	8 20	....	....	....	9 49	....	....	....	....	....	....
Pomeroy..+	42.3	....	....	8 28	....	....	....	A M	....	....	....	....	....	....
Parkesburg..+	44.3	....	....	8 34	....	....	....	....	....	....	....	....	....	....
(ARRIVE	....	....	....	A M	....	....	....	....	....	....	....	....	....	....

May 28, 1916.	1	5333	5339	5505	5507	5341	5345	5347	5351	5355	5357	5359	5361	5515	5366
Philada....lve.	A M	A M	Noon	Noon	Noon	PM	P M	P M	P M	P M	P M	▲	P M	P M	▲
Broad St...	*1130	*1145	*1215	k1240	♭1245	†115	*145	*215	*245	*315	†345	♭413	*415	†440	♭443
West Phila. ⊡	11 34	11 49	1219	12 44	12 49	1 19	1 49	2 19	2 49	3 19	3 49	4 17	4 19	4 44	4 47
Fortieth St.	- -	s1152	- -	- -	- -	- -	- -	- -	- -	- -	3 52	- -	- -	- -	- -
52d Street..	- -	11 54	1224	- -	12 55	1 24	1 54	2 24	2 54	3 24	3 55	- -	4 24	- -	- -
Overbrook..+	- -	11 57	1227	- -	12 59	1 27	1 57	2 27	2 57	3 27	3 58	-	4 27	- -	- -
Merion.....+	- -	11 59	1229	- -	1 01	1 29	1 59	2 29	2 59	3 29	4 00	- -	4 29	- -	- -
Narberth...+	- -	12 02	1232	- -	1 04	1 32	2 02	2 32	3 02	3 32	4 03	- -	4 32	- -	- -
Wynnewood+	- -	12 04	1234	- -	1 06	1 34	2 04	2 34	3 04	3 34	4 05	- -	4 34	- -	- -
Ardmore...+	- -	12 06	1236	- -	1 09	1 36	2 06	2 36	3 06	3 36	4 07	4 29	4 36	- -	4 59
Haverford...+	- -	12 08	1238	- -	1 12	1 38	2 08	2 38	3 08	3 38	4 09	4 31	4 38	- -	5 01
Bryn Mawr.+	- -	12 11	1241	- -	1 15	1 41	2 11	2 41	3 11	3 41	4 12	4 34	4 41	- -	5 04
Rosemont..+	- -	12 13	1243	1 02	1 17	1 43	2 13	2 43	3 13	3 43	4 14	4 36	4 43	- -	5 06
Villa Nova.+	- -	12 15	1245	1 05	1 20	1 45	2 15	2 45	3 15	3 45	4 16	4 38	4 45	- -	5 08
Radnor.....+	- -	12 18	1248	1 08	1 23	1 48	2 18	2 48	3 18	3 48	4 19	4 41	4 48	- -	5 11
St. David's.+	- -	12 20	1250	1 10	1 25	1 50	2 20	2 50	3 20	3 50	4 21	4 43	4 50	- -	5 13
Wayne.....+	- -	12 22	1252	1 12	1 27	1 52	2 22	2 52	3 22	3 52	4 23	4 45	4 52	- -	5 15
Strafford....+	- -	12 24	1254	1 15	1 30	1 54	2 24	2 54	3 24	3 54	4 25	4 47	4 54	- -	5 17
Devon......+	- -	12 27	1257	1 18	1 33	1 57	2 27	2 57	3 27	3 57	4 28	4 50	4 57	5 11	5 20
Berwyn....+	- -	12 29	1259	1 21	1 36	1 59	2 29	2 59	3 29	3 59	4 30	4 52	4 59	5 14	5 22
Daylesford...	- -	12 31	1 01	1 24	1 39	2 01	2 31	3 01	3 31	4 01	4 32	4 54	5 01	5 17	5 24
Paoli.......+	12 14	12 34	1 04	1 27	1 42	2 04	2 34	3 04	3 34	*411	4 35	4 57	5 04	5 20	5 27
Green Tree.+	- .-	Noon	P M	1 30	1 45	PM	P M	P M	P M	†414	P M	P M	P M	5 23	P M
Malvern....+	- -	....	....	1 32	1 47	....	....	....	....	4 16	....	....	....	5 25	....
Frazer...arr.	*1222	....	....	1 37	1 52	....	....	....	....	4 21		....	....	5 30	....
Frazer+.lve.	†1228	....	....	1 42	1 57	....	....	....	....	4 21		....	....	....	....
Villa Maria. (W. O. Branch)	12 30	....	....	1 44	1 59	....	....	....	....	4 23		....	....	....	....
Morstein. +	12 33	....	....	1 47	2 02	....	....	....	....	4 26		....	....	....	....
Kirkland...	12 36	....	....	1 50	2 05	....	....	....	....	4 29		....	....	....	....
Green Hill..+	12 40	....	....	1 54	2 09	....	....	....	....	4 32		....	....	....	....
Fern Hill...	12 43	....	....	1 57	2 12	....	....	....	....	4 35		....	....	....	....
Maple Ave..	12 46	....	....	2 00	2 15	....	....	....	....	4 37		....	....	....	....
W. Chester+ar.	†1248	....	....	2 02	2 17	....	....	....	....	†439		....	....	....	....
Glen Loch..+	- -	....	....	1 40	1 55	....	....	....	....	P M		....	....	5 33	....
Ship Road....	- -	....	....	1 43	1 58	....	....	....	....	....		....	....	5 36	....
Whiteland....	- -	....	....	1 46	2 01	....	....	....	....	....		....	....	5 38	....
Whitford...+	- -	....	....	1 49	2 04	....	....	....	....	....		....	....	5 41	....
Bradford Hills	- -	....	....	1 52	2 07	....	....	....	....	....		....	....	5 44	....
Woodbine..+	- -	....	....	1 55	2 10	....	....	....	....	....		....	....	5 47	....
Downingtown.+	12 34	....	....	1 59	2 14	....	....	....	....	....		....	....	5 50	....
Thorndale..+	- -	....	....	P M	P M	....	....	....	....	....		....	....	5 55	....
Cain.........	- -	....	....	....	....	....	....	....	....	....		....	....	5 58	....
Coatesville.+	12 46	....	....	....	....	....	....	....	....	....		....	....	6 03	....
Pomeroy..+	- -	....	....	....	....	....	....	....	....	....		....	....	6 09	....
Parkesburg..+	*1257	....	....	....	....	....	....	....	....	....		....	....	6 14	....
(ARRIVE	Noon	....	....	....	....	....	....	....	....	....		....	....	P M	....

Additional Trains—Leave Philadelphia for Paoli k12 45 noon, k1 05, k1 35, k2 05 p.m.

Eastern time.)	5367	5373	5377	5517	5379	5381	5383	5385	5387	5389	5391	5393	5395	5401
Philada....lve.	P M	▲	P M	P M	▲	P M	▲	PM	P M	PM	P M	PM	P M	P M
Broad Street.	*445	♭510	†5 15	†530	♭540	*545	†606	*615	*645	*715	†745	*815	†845	†915
West Phila. ⊡	4 49	5 14	5 19	5 34	5 44	5 49	6 09	6 19	6 49	7 19	7 49	8 19	8 49	9 19
Fortieth St...	- -	- -	- -	- -	- -	- -	- -	- -	- -	- -	- -	8 22	- -	- -
52d Street....	4 54	- -	5 24	- -	- -	5 54	- -	6 24	6 54	7 24	7 54	8 25	8 54	9 24
Overbrook......	4 57	- -	5 27	- -	- -	5 57	- -	6 27	6 57	7 27	7 57	8 28	8 57	9 27
Merion.........	4 59	- -	5 29	- -	- -	5 59	- -	6 29	6 59	7 29	7 59	8 30	8 59	9 29
Narberth.......	5 02	- -	5 32	- -	- -	6 02	- -	6 32	7 02	7 32	8 02	8 33	9 02	9 32
Wynnewood....	5 04	- -	5 34	- -	- -	6 04	- -	6 34	7 04	7 34	8 04	8 35	9 04	9 34
Ardmore.......	5 06	5 26	5 36	- -	5 56	6 06	6 21	6 36	7 06	7 36	8 06	8 37	9 06	9 36
Haverford......	5 08	5 28	5 38	- -	5 58	6 08	6 23	6 38	7 08	7 38	8 08	8 39	9 08	9 38
Bryn Mawr.....	5 11	5 31	5 41	- -	6 01	6 11	6 25	6 41	7 11	7 41	8 11	8 42	9 11	9 41
Rosemont......	5 13	5 33	5 43	5 52	6 03	6 13	6 27	6 43	7 13	7 43	8 13	8 44	9 13	9 43
Villa Nova.....	5 15	5 35	5 45	5 55	6 05	6 15	6 30	6 45	7 15	7 45	8 15	8 46	9 15	9 45
Radnor.........	5 18	5 38	5 48	5 58	6 08	6 18	6 32	6 48	7 18	7 48	8 18	8 49	9 18	9 48
St. David's.....	5 20	5 40	5 50	6 00	6 10	6 20	6 34	6 50	7 20	7 50	8 20	8 51	9 20	9 50
Wayne.........	5 22	5 42	5 52	6 03	6 12	6 22	6 36	6 52	7 22	7 52	8 22	8 53	9 22	9 52
Strafford.......	5 24	5 44	5 54	6 06	6 14	6 24	6 39	6 54	7 24	7 54	8 24	8 55	9 24	9 54
Devon..........	5 27	5 47	5 57	6 09	6 17	6 27	6 41	6 57	7 27	7 57	8 27	8 58	9 27	9 57
Berwyn........	5 29	5 49	5 59	6 12	6 19	6 29	6 44	6 59	7 29	7 59	8 29	9 00	9 29	9 59
Daylesford.....	5 31	5 51	6 01	6 15	6 21	6 31	6 46	7 01	7 31	8 01	8 31	9 02	9 31	1001
Paoli...........	5 34	5 58	6 04	6 18	6 24	*534	6 49	7 04	*758	8 04	8 34	9 05	9 38	1004
Green Tree.....	P M	6 01	k611	6 21	P M	§641	P M	PM	†741	PM	P M	PM	9 41	P M
Malvern........	....	6 03	k613	6 23	....	6 43	....	....	7 43	....	....	....	9 43	....
Frazer.....arr.	....	6 08	k618	6 28	....	6 48	....	....	7 48	....	....	....	9 48	....
Frazer.....lve.	....	6 08	k618	....	....	6 48	....	....	7 48	....	....	....	9 48	....
Villa Maria.. (W. O. Branch)	....	6 10	k620	....	....	6 50	....	....	7 50	....	....	....	9 50	....
Morstein......	....	6 13	k623	....	....	6 53	....	....	7 53	....	....	....	9 53	....
Kirkland......	....	6 16	k626	....	....	6 56	....	....	7 56	....	....	....	9 56	....
Green Hill.....	....	6 19	k629	....	....	6 59	....	....	7 59	....	....	....	9 59	....
Fern Hill......	....	6 22	k632	....	....	7 02	....	....	8 02	....	....	....	1002	....
Maple Ave.....	....	6 24	k634	....	....	7 04	....	....	8 04	....	....	....	1004	....
W. Chester. .arr.	....	6 26	k636	....	....	§706	....	....	†806	....	....	....	1005	....
Glen Loch......	....	PM	P M	6 31	....	P M	....	....	P M	....	....	....	P M	....
Ship Road......	....	....	....	6 34	....	....	....	....	....	....	....	....	....	....
Whiteland......	....	....	....	6 36	....	....	....	....	....	....	....	....	....	....
Whitford.......	....	....	....	6 39										
Bradford Hills..	....	....	....	6 42										
Woodbine......	....	....	....	6 45										
Downingtown	....	....	....	6 48										
Thorndale......	....	....	....	6 53										
Caln...........	....	....	....	6 56										
Coatesville.....	....	....	....	7 01										
Pomeroy.......	....	....	....	7 07										
Parkesburg..	....	....	....	7 12										
(ARRIVE	....	....	....	P M										

Eastern time.)	5405	5409	5411	5519	5301	643	5709	5727	5825	5829	5739	5745	5833	5769	5773
Philada....lve.	P M	P M	P M	†PM	Night	A M	AM	§N'n	PM	P M	P M	P M	P M	§PM	P M
Broad Street.	*945	*1030	†1110	1130	†1215	§710	§815	1245	§115	§240	§345	§515	§915	1100	§1145
West Phila. ⊡	9 49	10 34	1114	1134	1219	7 14	8 19	1249	1 19	2 44	3 49	5 19	9 19	1104	1149
Fortieth St...	- -	- -	- -	1138	- -	- -	8 22	- -	- -	- -	- -	- -	9 23	- -	- -
52d Street....	9 54	10 39	1119	1142	1224	- -	8 25	1254	1 25	2 50	3 54	5 24	9 27	1109	1154
Overbrook......	9 57	10 42	1122	1146	1227	- -	8 28	1257	1 29	- -	3 57	5 27	9 31	1112	1157
Merion.........	9 59	10 44	1124	1148	1229	- -	8 30	1259	1 31	- -	3 59	5 29	9 33	1114	1159
Narberth.......	1002	10 47	1127	1151	1232	- -	8 33	1 02	1 34	- -	4 02	5 32	9 36	1117	1202
Wynnewood....	1004	10 49	1129	1153	1234	- -	8 35	1 04	1 36	- -	4 04	5 34	9 38	1119	1204
Ardmore.......	1006	10 51	1131	1156	1236	- -	8 37	1 06	1 39	- -	4 06	5 36	9 41	1121	1206
Haverford......	1008	10 53	1133	1159	1238	- -	8 39	1 08	1 42	- -	4 08	5 38	9 44	1123	1208
Bryn Mawr.....	1011	10 56	1136	1202	1241	- -	8 42	1 11	1 45	- -	4 11	5 41	9 47	1126	1211
Rosemont......	1013	10 58	1138	1204	1243	- -	8 44	1 13	1 47	- -	4 13	5 43	9 49	1128	1213
Villa Nova.....	1015	11 00	1140	1207	1245	- -	8 46	1 16	1 50	- -	4 15	5 45	9 52	1130	1215
Radnor.........	1018	11 03	1143	1210	1248	- -	8 49	1 18	1 53	- -	4 18	5 48	9 55	1133	1218
St. David's.....	1020	11 05	1145	1212	1250	- -	8 51	1 20	1 55	- -	4 20	5 50	9 57	1135	1220
Wayne.........	1022	11 07	1147	1214	1252	- -	8 53	1 22	1 57	- -	4 22	5 52	9 59	1137	1222
Strafford.......	1024	11 09	1149	1217	1254	- -	8 55	1 24	2 00	- -	4 24	5 54	1002	1139	1224
Devon..........	1027	11 12	1152	1220	1257	- -	8 58	1 27	2 03	- -	4 27	5 57	1005	1142	1227
Berwyn........	1029	11 14	1154	1223	1258	- -	9 00	1 29	2 06	- -	4 29	5 59	1008	1144	1229
Daylesford.....	1031	11 16	1156	1226	1 01	- -	9 02	1 31	2 09	- -	4 31	6 01	1011	1146	1231
Paoli...........	1034	11 19	1203	1229	¶104	7 47	9 12	1 38	2 12	3 15	4 38	6 04	1014	1153	1234
Green Tree.....	P M	P M	1206	1232	A M	7 50	9 15	1 41	2 15	3 18	4 41	P M	1017	1156	Night
Malvern........	....	....	1208	1234	....	7 53	9 17	1 43	2 17	3 20	4 43	....	1019	1158	....
Frazer.....arr.	....	....	1213	1239	....	7 59	9 22	1 48	2 22	3 25	4 48	....	1024	1203	....
Frazer.....lve.	....	....	1213	....	....	....	9 22	1 48	....	....	4 48	....	....	1203	....
Villa Maria.. (W. O. Branch)	....	....	1215	....	....	....	9 24	1 50	....	....	4 50	....	....	1205	....
Morstein......	....	....	1218	....	....	....	9 27	1 53	....	....	4 53	....	....	1208	....
Kirkland......	....	....	1221	....	....	....	9 30	1 56	....	....	4 55	....	....	1211	....
Green Hill.....	....	....	1224	....	....	....	9 33	1 59	....	....	4 59	....	....	1214	....
Fern Hill......	....	....	1227	....	....	....	9 36	2 02	....	....	5 02	....	....	1217	....
Maple Ave.....	....	....	1229	....	....	....	9 38	2 04	....	....	5 04	....	....	1219	....
W. Chester. .arr.	....	....	1231	....	....	....	9 40	2 06	....	....	5 06	....	....	1221	....
Glen Loch......	....	....	Night	1242	....	8 02	AM	P M	2 25	3 28	P M	....	1027	Night	....
Ship Road......	....	....	....	1244	....	8 05	....	....	2 28	3 31	....	....	1030	....	....
Whiteland......	....	....	....	1246	....	8 08	....	....	2 30	3 34	....	....	1032	....	....
Whitford.......				1248	....	8 11	....	....	2 33	3 37	....	....	1035	....	....
Bradford Hills..				1251	....	8 14	....	....	2 36	3 40	....	....	1038	....	....
Woodbine......				1254	....	8 17	....	....	2 39	3 43	....	....	1041	....	....
Downingtown				1257	....	8 21	....	....	2 42	3 46	....	....	1044	....	....
Thorndale......				Night	....	8 26	....	....	PM	3 51	....	....	P M	....	....
Caln...........				....	....	8 30	....	....	....	3 54	....	....	....	....	....
Coatesville.....				....	....	8 36	....	....	....	3 59	....	....	....	....	....
Pomeroy.......				....	....	8 43	....	....	....	4 06	....	....	....	....	....
Parkesburg..				....	....	8 48	....	....	....	4 11	....	....	....	....	....
(ARRIVE				....	....	A M	....	....	....	P M	....	....	....	....	....

Additional Sunday Train—Lvs. Philadelphia (Broad Street) 6 45 a.m., arr. Paoli 7 34 a.m., making all stops.

* Daily; † daily, except Sunday; ¶ daily, except Monday; § Sunday only; c regular stop to take on passengers; k Saturday only; s stops on Sunday; t does not stop on Sunday. ♭ Daily, except Saturday and Sunday. ▲ Will not run holidays. • Via connecting train. ⊡ Westbound trains make a regular stop at West Philadelphia only to take on passengers. + Coupon stations.

The Pennsylvania Railroad Table 29 Philadelphia Division May 28, 1916 schedule from the *June 1916 Official Guide of the Railways* on the 30.8-mile trip from Philadelphia via Frazer to West Chester showed thirteen local passenger trains on Monday to Saturday and nine trips on Sunday. Depending on the number of stops the one-way trip could require from one hour and ten minutes to one hour and thirty minutes. As automobiles became more popular, ridership declined, and the last passenger train operated via Frazer to West Chester on September 24, 1932.

438

PENNSYLVANIA SYSTEM

Adams Express Company

PENNSYLVANIA RAILROAD

Table 29—PHILADELPHIA DIVISION.—Local Trains Parkesburg and West Chester to Philadelphia.
For Additional Trains between West Chester and Philadelphia, see P. B. & W. R.R. (Media Division.)

May 28, 1916.	5302	5304	5472	5308	5500	5312	5314	5504	5318	5320	5322	5508	5324	5326	5604	5328	5476	5336	5338	5478	5344	5480	5350	5352	5356	5512	5358	5360	5364	5370
LEAVE					AM							AM			AM											Noon				
Parkesburg					†6 25							†7 13			†8 21											†12 33				
Pomeroy					6 29							7 18			8 25											12 40				
Coatesville					6 36							7 27			8 33											12 48				
Cain					6 40							7 31			– –											12 53				
Thorndale					6 43			AM				7 36			– –											12 57				
Downingtown					6 48			†7 17				7 43			8 42											1 03				
Woodbine					6 51			7 20				7 46			– –											1 06				
Bradford Hill					6 54			7 23				7 49			– –											1 10				
Whitford					6 57			7 26				7 53			– –											1 15				
Whiteland					7 00			7 29				7 56			– –											1 18				
Ship Road					7 03			7 32				7 59			– –											1 21				
Glen Loch			AM		7 06			7 35				8 03			– –		AM			AM	AM	AM			Noon	1 24				
West Chester, lv.			†6 17									†7 42					†8 43			†10 13	§10 43	†11 57			§12 43	†1 07				
Maple Avenue			6 19									7 44					8 45			10 15	10 45	11 59			12 45	1 09				
Fern Hill			6 21									7 47					8 47			10 17	10 47	12 01			12 47	1 11				
Green Hill			6 24									7 50					8 50			10 20	10 50	12 04			12 50	1 14				
Kirkland			6 27									7 54					8 53			10 23	10 53	12 07			12 53	1 17				
Morstein			6 30									7 57					8 56			10 26	10 56	12 10			12 56	1 20				
Villa Maria			6 33									8 00					8 59			10 29	10 59	12 13			12 59	1 23				
Frazer, arr.			6 35									8 02					9 01			10 31	11 01	12 15			1 01	1 25				
Frazer, lve.			6 35		7 10			7 39				8 07			8 56		9 01			10 31	11 01				1 01	1 29				
Malvern			6 40	▲	7 15	▲	▲	7 44	▲	▲		8 13			– –		9 06			10 36	11 06	5346		Noon	1 06	1 35	▲			
Green Tree	AM	AM	6 42	AM	7 17	AM	AM	7 46	AM	AM	AM	8 16	AM	AM	– –	AM	9 08	AM	AM	10 38	§11 08	AM	Noon	†	§1 08	1 37	PM	PM	PM	PM
Paoli	†5 58	†6 28	6 58	†7 12	7 21	†7 28	†7 44	7 50	†7 58	†8 10	†8 16	8 20	†8 28	†8 45	9 04	†8 58	*9 15	*9 46	*10 16	10 46	*11 16	†11 46	*12 16	12 46	*1 16	1 40	†1 46	†2 16	*2 46	*3 16
Daylesford	6 00	6 30	7 00	7 14	7 23	7 30	7 46	7 52	8 00	8 12	– –	– –	8 30	8 47	– –	9 00	9 18	9 48	10 18	10 48	11 18	11 48	12 18	12 48	1 18	– –	1 48	2 18	2 48	3 18
Berwyn	6 02	6 32	7 02	7 16	7 26	7 32	7 48	7 55	8 02	8 14	8 20	b8 24	8 32	8 49	– –	9 02	9 20	9 50	10 20	10 50	11 20	11 50	12 20	12 50	1 20	– –	1 50	2 20	2 50	3 20
Devon	6 05	6 35	7 05	7 19	7 29	7 35	7 51	7 58	8 05	8 17	8 22	– –	8 35	8 52	– –	9 06	9 23	9 53	10 23	10 53	11 23	11 53	12 23	12 53	1 23	– –	1 53	2 23	2 53	3 23
Strafford	6 07	6 37	7 07	7 21	7 31	7 37	7 53	w –	8 07	8 19	8 24	– –	8 37	8 54	– –	9 07	9 25	9 55	10 25	10 55	11 25	11 55	12 25	12 55	1 25	– –	1 55	2 25	2 55	3 25
Wayne	6 09	6 39	7 09	7 23	7 34	7 39	7 55	8 01	8 03	8 21	8 26	b8 32	8 39	8 56	– –	9 09	9 27	9 57	10 27	10 57	11 27	11 57	12 27	12 57	1 27	– –	1 57	2 27	2 57	3 27
St. David's	6 11	6 41	7 11	7 25	7 36	7 41	7 57	8 03	8 11	8 23	8 28	– –	8 41	8 58	– –	9 11	9 29	9 59	10 29	10 59	11 29	11 59	12 29	12 59	1 29	– –	1 59	2 29	2 59	3 29
Radnor	6 13	6 43	7 13	7 27	7 38	7 43	7 59	8 05	8 13	– –	8 30	– –	8 43	9 00	– –	9 13	9 31	10 01	10 31	11 01	11 31	12 01	12 31	1 01	1 31	– –	2 01	2 31	3 01	3 31
Villa Nova	6 15	6 45	7 15	7 29	w –	7 45	8 01	w –	8 15	– –	8 32	– –	8 45	9 02	– –	9 15	9 33	10 03	10 33	11 03	11 33	12 03	12 33	1 03	1 33	– –	2 03	2 33	3 03	3 33
Rosemont	6 17	6 47	7 17	7 31	w –	7 47	8 03	8 09	8 17	– –	8 34	– –	8 47	9 04	– –	9 17	9 35	10 05	10 35	11 05	11 35	12 05	12 35	1 05	1 35	– –	2 05	2 35	3 05	3 35
Bryn Mawr	6 20	6 50	7 20	7 34	w –	7 50	8 06	8 12	8 20	– –	8 37	b8 40	8 50	9 07	– –	9 20	9 38	10 08	10 38	11 08	11 38	12 08	12 38	1 08	1 38	– –	2 08	2 38	3 08	3 38
Haverford	6 22	6 52	7 22	7 36	w –	7 52	8 08	8 15	8 22	8 32	8 39	b8 43	8 52	9 09	– –	9 22	9 40	10 10	10 40	11 10	11 40	12 10	12 40	1 10	1 40	– –	2 10	2 40	3 10	3 40
Ardmore	6 24	6 54	7 24	7 38	w –	7 54	8 10	8 17	8 24	– –	8 42	b8 45	8 54	9 11	– –	9 24	9 42	10 12	10 42	11 12	11 42	12 12	12 42	1 12	1 42	– –	2 12	2 42	3 12	3 42
Wynnewood	6 27	6 57	7 27	7 41	w –	7 57	– –	8 19	8 27	– –	– –	– –	8 57	– –	– –	9 27	9 45	10 15	10 45	11 15	11 45	12 15	12 45	1 15	1 45	– –	2 15	2 45	3 15	3 45
Narberth	6 29	6 59	7 29	7 43	w –	7 59	8 13	w –	8 29	– –	– –	– –	8 59	– –	– –	9 29	9 47	10 17	10 47	11 17	11 47	12 17	12 47	1 17	1 47	– –	2 17	2 47	3 17	3 47
Merion	6 31	7 01	7 31	– –	w –	8 01	– –	w –	8 31	– –	– –	– –	9 01	– –	– –	9 31	9 49	10 19	10 49	11 19	11 49	12 19	12 49	1 19	1 49	– –	2 19	2 49	3 19	3 49
Overbrook	6 33	7 03	7 33	– –	w –	8 03	8 17	w –	8 33	– –	– –	– –	9 03	– –	– –	9 33	9 51	10 21	10 51	11 21	11 51	12 21	12 51	1 21	1 51	– –	2 21	2 51	3 21	3 51
52d Street	6 36	7 06	7 36	– –	w –	8 06	– –	w –	8 36	– –	– –	– –	9 06	– –	– –	9 36	9 54	10 24	10 54	11 24	11 54	12 24	12 54	1 24	1 54	– –	2 24	2 54	3 24	3 54
Fortieth St.	– –	– –	– –	– –	7 54	– –	– –	– –	– –	– –	– –	– –	– –	– –	– –	– –	– –	s –	– –	– –	– –	– –	– –	– –	– –	– –	– –	– –	– –	– –
West Phila ▣	6 41	7 11	7 41	7 52	7 58	8 11	8 24	8 31	8 41	8 44	8 53	8 56	9 11	9 22	9 31	9 41	9 59	10 29	10 59	11 29	11 59	12 29	12 59	1 29	1 59	2 06	2 29	2 59	3 29	3 59
Broad Street	6 45	7 15	7 45	7 56	8 02	8 15	8 28	8 35	8 45	8 48	8 57	9 00	9 15	9 26	9 35	9 45	10 03	10 33	11 03	11 33	12 03	12 33	1 03	1 33	2 05	2 10	2 33	3 03	3 33	4 03
Philada, arr.	AM	AM	AM	AM	AM	AM	AM	AM	AM	AM	AM	AM	AM	AM	AM	AM	AM	AM	AM	AM	AM	Noon	Noon	PM	PM	PM	PM	PM	PM	PM

(Eastern time.)	5372	5374	5378	5380	5382	5388	5516	5390	5392	5394	5398	5400	5518	5402	5406	604	5492	5300	5702	5704	5705	5708	5824	5828	5832	5816	5746	5818	1072
LEAVE													PM			PM							AM						PM
Parkesburg													†7 53			†9 26							§7 54						§9 09
Pomeroy													7 58			9 30							7 59						9 13
Coatesville													8 06			9 38							8 07						9 20
Cain													8 10			9 41							8 11						9 24
Thorndale							PM						8 14			9 46							8 15	PM	PM				9 27
Downingtown							†5 52						8 19			9 53							8 22	§1 45	§3 45				9 32
Woodbine							5 55						8 22			9 56							8 25	1 48	3 48				9 35
Bradford Hill							5 58						8 25			9 59							8 28	1 51	3 51				9 38
Whitford							6 01						8 29			10 02							8 31	1 54	3 54				9 41
Whiteland							6 04						8 32			10 05							8 34	1 57	3 57				9 44
Ship Road							6 07						8 35			10 07							8 37	1 59	3 59				9 47
Glen Loch	PM				PM		6 10		PM				8 39	PM		10 10	PM					AM	8 41	2 02	4 02	PM		PM	9 50
West Chester, lv.	*3 13				†5 13				†6 46					†8 55			†10 40					§8 13	§8 13			§5 43		§9 25	§9 25
Maple Avenue	3 15				5 15				6 48					8 57			10 42					8 15	8 15			5 45		9 27	9 27
Fern Hill	3 17				5 17				6 50					8 59			10 44					8 17	8 17			5 47		9 29	9 29
Green Hill	3 20				5 20				6 53					9 02			10 47					8 20	8 20			5 50		9 32	9 32
Kirkland	3 23				5 23				6 56					9 05			10 50					8 23	8 23			5 53		9 35	9 35
Morstein	3 26				5 26				6 59					9 08			10 53					8 26	8 26			5 56		9 38	9 38
Villa Maria	3 29				5 29				7 02					9 11			10 56					8 29	8 29			5 59		9 41	9 41
Frazer, arr.	3 31				5 31				7 04					9 13			10 58					8 31	8 31			6 01		9 43	9 43
Frazer, lve.	3 31				5 31		6 13		7 04				8 43	9 13		10 14	10 58					8 31	8 45	2 05	4 05	6 01		9 43	9 54
Malvern	3 36				5 36		6 20		7 09				8 49	9 18		10 19	11 03					8 36	8 50	2 10	4 10	6 06		9 48	10 00
Green Tree	3 38	PM	PM	PM	5 38	PM	6 22	PM	7 11	PM	PM	PM	8 51	9 20	PM	10 21	†11 08	Night	AM	AM	AM	8 38	8 52	2 12	4 12	6 08	PM	9 50	10 02
Paoli	3 46	†4 16	*4 46	*5 16	*5 46	†6 16	6 25	†6 58	7 19	*7 28	*7 58	*8 28	*8 55	*9 28	†9 58	*10 25	*11 16	†12 18	§6 46	§7 20	§8 16	8 46	8 55	2 15	4 15	6 11	§6 46	9 58	10 05
Daylesford	3 48	4 18	4 48	5 18	5 48	6 18	6 27	7 00	7 21	7 30	8 00	8 30	8 57	9 30	10 00	10 27	11 20	12 20	6 48	7 22	8 18	8 48	– –	2 17	4 17	6 16	6 48	10 00	– –
Berwyn	3 50	4 20	4 50	5 20	5 50	6 20	6 30	7 02	7 23	7 32	8 02	8 32	9 00	9 32	10 02	10 30	11 22	12 22	6 50	7 24	8 20	8 50	– –	2 20	4 20	6 18	6 50	10 02	– –
Devon	3 53	4 23	4 53	5 23	5 53	6 23	6 33	7 05	7 26	7 35	8 05	8 35	9 03	9 35	10 05	10 33	11 25	12 25	6 53	7 27	8 23	8 53	– –	2 23	4 23	6 20	6 53	10 05	– –
Strafford	3 55	4 25	4 55	5 25	5 55	6 25	6 35	7 07	7 28	7 37	8 07	8 37	9 05	9 37	10 07	10 35	11 27	12 27	6 55	7 29	8 25	8 55	– –	2 25	4 25	6 23	6 55	10 07	– –
Wayne	3 57	4 27	4 57	5 27	5 57	6 27	6 38	7 09	7 30	7 39	8 09	8 39	9 08	9 39	10 09	10 38	11 29	12 29	6 57	7 31	8 27	8 57	– –	2 28	4 28	6 25	6 57	10 09	– –
St. David's	3 59	4 29	4 59	5 29	5 59	6 29	6 40	7 11	7 32	7 41	8 11	8 41	9 10	9 41	10 11	10 40	11 31	12 31	6 59	7 33	8 29	8 59	– –	2 30	4 30	6 27	6 59	10 11	– –
Radnor	4 01	4 31	5 01	5 31	6 01	6 31	6 42	7 13	7 34	7 43	8 13	8 43	9 12	9 43	10 13	10 42	11 33	12 33	7 01	7 35	8 31	9 01	– –	2 32	4 32	6 29	7 01	10 13	– –
Villa Nova	4 03	4 33	5 03	5 33	6 03	6 33	6 45	7 16	7 36	7 45	8 15	8 45	9 15	9 45	10 15	10 45	11 35	12 35	7 03	7 37	8 33	9 03	– –	2 35	4 35	6 31	7 03	10 15	– –
Rosemont	4 05	4 35	5 05	5 35	6 05	6 35	6 47	7 17	7 38	7 47	8 17	8 47	9 17	9 47	10 17	10 47	11 37	12 37	7 05	7 39	8 35	9 05	– –	2 37	4 37	6 33	7 05	10 17	– –
Bryn Mawr	4 08	4 38	5 08	5 38	6 08	6 38	6 50	7 20	7 41	7 50	8 20	8 50	9 20	9 50	10 20	10 50	11 40	12 40	7 08	7 42	8 38	9 08	– –	2 40	4 40	6 35	7 08	10 20	– –
Haverford	4 10	4 40	5 10	5 40	6 10	6 40	6 53	7 22	7 43	7 52	8 22	8 52	9 23	9 52	10 22	10 53	11 42	12 42	7 10	7 44	8 40	9 10	– –	2 43	4 43	6 38	7 10	10 22	– –
Ardmore	4 12	4 42	5 12	5 42	6 12	6 42	6 55	7 24	7 45	7 54	8 24	8 54	9 25	9 54	10 24	10 55	11 44	12 44	7 12	7 46	8 42	9 12	– –	2 45	4 45	6 40	7 12	10 24	– –
Wynnewood	4 15	4 45	5 15	5 45	6 15	6 45	6 57	7 27	– –	7 57	8 27	8 57	9 27	9 57	10 27	10 57	11 47	12 47	7 15	7 49	8 45	9 15	– –	2 47	4 47	6 42	7 15	10 27	– –
Narberth	4 17	4 47	5 17	5 47	6 17	6 47	6 59	7 29	– –	7 59	8 29	8 59	9 29	9 59	10 29	10 59	11 49	12 49	7 17	7 51	8 47	9 17	– –	2 49	4 49	6 45	7 17	10 29	– –
Merion	4 19	4 49	5 19	5 49	6 19	6 49	7 01	7 31	– –	8 01	8 31	9 01	9 31	10 01	10 31	11 01	11 51	12 51	7 19	7 53	8 49	9 19	– –	2 51	4 51	6 47	7 19	10 31	– –
Overbrook	4 21	4 51	5 21	5 51	6 21	6 51	7 03	7 33	– –	8 03	8 33	9 03	9 33	10 03	10 33	11 03	11 53	12 53	7 21	7 55	8 51	9 21	– –	2 53	4 53	6 49	7 21	10 33	– –
52d Street	4 24	4 54	5 24	5 54	6 24	6 54	7 06	7 36	– –	8 06	8 36	9 06	9 36	10 06	10 36	11 06	11 56	12 56	7 24	7 58	8 54	9 24	– –	2 56	4 56	6 51	7 24	10 36	– –
Fortieth St.	– –	– –	– –	– –	s –	– –	– –	– –	– –	– –	– –	– –	– –	s –	– –	– –	– –	– –	– –	– –	– –	– –	– –	– –	– –	– –	– –	– –	– –
West Phila ▣	4 29	4 59	5 29	5 59	6 29	6 59	7 11	7 41	7 56	8 11	8 41	9 11	9 41	10 11	10 41	11 11	12 01	1 01	7 29	8 03	8 59	9 29	9 21	3 01	5 01	6 59	7 29	10 41	10 31
Broad Street	4 33	5 03	5 33	6 03	6 33	7 03	7 15	7 45	8 00	8 15	8 45	9 15	9 45	10 15	10 45	11 15	12 05	¶1 05	7 33	8 07	9 03	9 33	9 25	3 05	5 05	7 03	7 33	10 45	10 35
Philada, arr.	PM	PM	PM	PM	PM	PM	PM	PM	PM	PM	PM	PM	PM	PM	PM	PM	Night	AM	AM	AM	AM	AM	AM	PM	PM	PM	PM	PM	PM

* Daily; † daily, except Sunday; ¶ daily, except Monday; § Sunday only.

▣ Eastbound trains make a regular stop at West Philadelphia only to leave passengers.

b Stops only on notice to conductor to leave passengers; s stops Sunday only; w stops holidays. ▲ Will not run holidays.

The May 28, 1916 Pennsylvania Railroad Table 29 schedule from the *June 1916 Official Guide of the Railways* showed eleven Monday to Saturday trains and six Sunday trains from West Chester via Frazer to Philadelphia. This provided nice accessibility for residents in the rural area particularly between West Chester and Frazier.

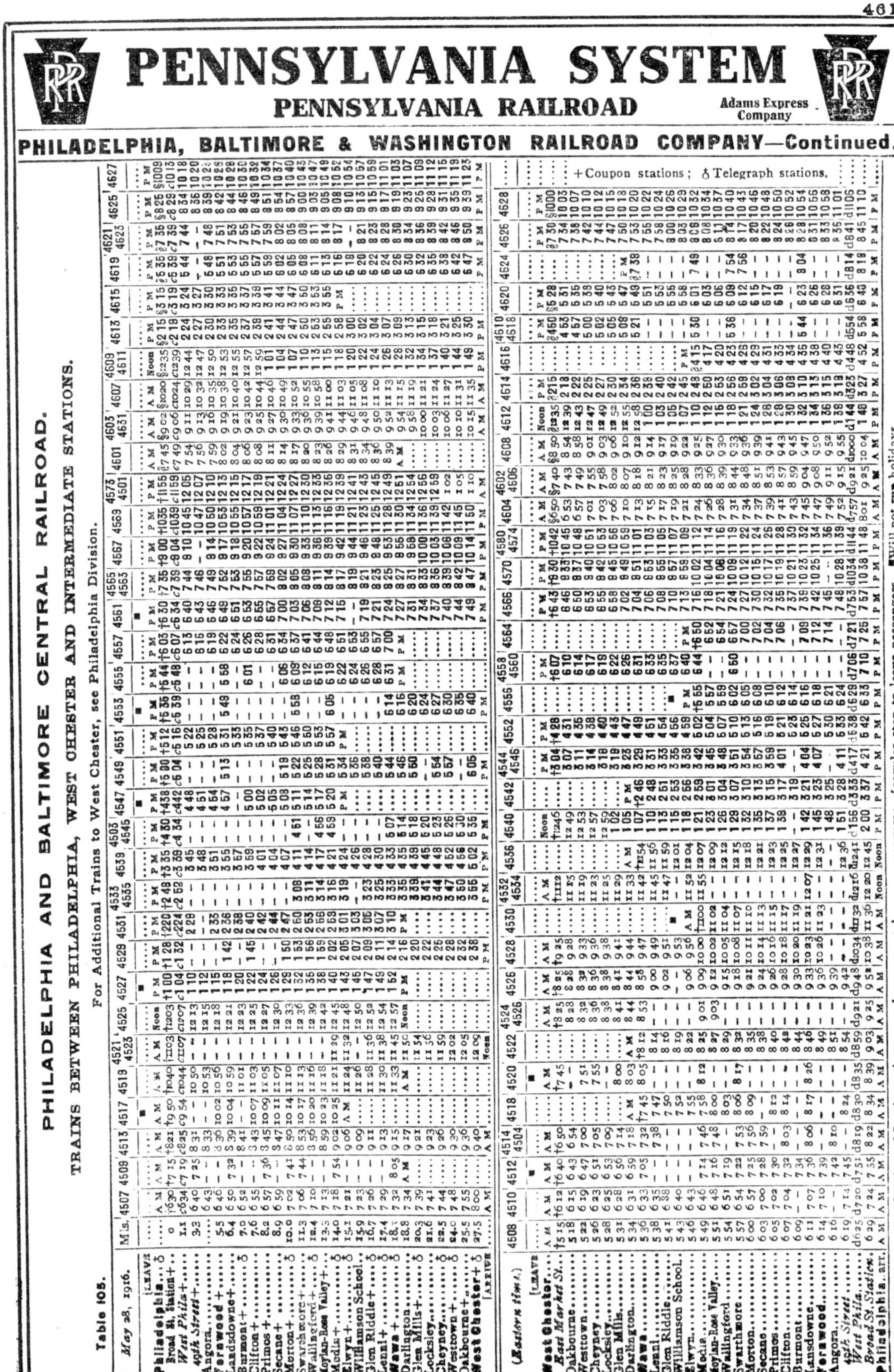

The Pennsylvania System (Philadelphia & Baltimore Central Railroad) Table 105 May 28, 1916 train schedule from the *June 1916 Official Guide of the Railways* between Philadelphia via Media to West Chester shows fourteen trains Monday to Saturday and eight trains on Sunday from Philadelphia to West Chester. From West Chester via Media to Philadelphia, there were sixteen trains Monday to Saturday and nine on Sunday. Typically, travel time from Philadelphia to West Chester via Media was about one hour and thirty minutes.

285

PENNSYLVANIA RAILROAD

Largest Fleet of Air-Conditioned Trains in the World

Table 97.

PHILADELPHIA DIVISION.

Eastern Standard time. Local Trains Philadelphia to Paoli—Bus Service Paoli to West Chester. Eastern Region

July 12, 1936.	Ms.	605	5301	Bus	5303	5305	Bus	5307	5309	5311	Bus
Philadelphia + ...lve.		A M			Ψ	◈■		■	◈■	⊙	
Broad Street	0	†2 30	A M		A M	A M		A M	A M	A M	
Broad St. Sub'n Sta.		– –	†4 55		*5 25	†6 15		*6 45	†7 15	*7 45	
Penna. Sta. (30th St.)		2 39	4 58		5 28	6 18		6 48	7 18	7 48	
52d Street	4.1	– –	5 04		5 33	6 23		6 53	7 23	7 54	
Overbrook +	5.6	– –	5 07		5 36	6 26		6 56	7 26	7 57	
Merion +	6.1	– –	5 09		5 38	6 28		6 58	7 28	7 59	
Narberth +	7.0	2 52	5 11		5 40	6 30		7 00	7 30	8 01	
Wynnewood +	7.6	– –	5 13		5 42	6 32		7 02	7 32	8 03	
Ardmore +	8.6	3 01	5 17		5 44	6 34		7 04	7 34	8 05	
Haverford +	9.3	3 04	5 20		5 46	6 36		7 06	7 36	8 07	
Bryn Mawr +	10.3	3 09	5 24		5 48	6 38		7 08	7 38	8 09	
Rosemont +	11.0	– –	5 26		5 50	6 40		7 10	7 40	8 11	
Villanova +	12.1	– –	5 29		5 52	6 42		7 12	7 42	8 13	
Radnor +	13.1	– –	5 32		5 54	6 44		7 14	7 44	8 15	
St. Davids +	13.9	f3 16	5 34		5 56	6 46		7 16	7 46	8 17	
Wayne +	14.6	3 18	5 36		5 58	6 48		7 18	7 48	8 19	
Strafford +	15.5	– –	5 39		6 00	6 50		7 20	7 50	8 21	
Devon +	16.6	– –	5 42		6 02	6 52		7 22	7 52	8 23	
Berwyn +	17.7	3 29	5 45		6 04	6 54		7 24	7 54	8 25	
Daylesford	18.7	– –	– –		f6 06	f6 56		f7 26	– –	f8 27	
Paoli +	20.0	3 35	5 52		6 09	6 59		7 29	7 59	8 30	
[ARRIVE			A M		A M	A M		A M	A M	A M	
				☽◈			☽◈				☽
[LEAVE				A M			A M				A M
Paoli	20.0	†3 35		†5 57			†7 10				k8 35
Green Tree	21.1	– –		6 00			7 13				8 38
Malvern +	21.7	3 40		6 03			7 16				8 41
Frazer...(*P.R.R. Sta.*)	24.0	A M									
Frazer (King & Goshen Rd.)	24.0			6 09			7 22				8 47
Immaculata	24.7			6 12			7 25				8 50
Morstein	25.7			6 14			7 27				8 52
Kirkland	27.0			6 17			7 30				8 55
Green Hill	28.1			6 19			7 32				8 57
Fern Hill	29.4			6 23			7 36				9 01
Biddle Street	30.3			6 27			7 40				9 05
West Chester +	30.8			6 32			7 45				9 10
[ARRIVE				A M			A M				A M

July 12, 1936.	5313	5703	5315	5317	5319	5321	5323	5451	5325	5327	5329	5331	5333	5335	5337
Philadelphia + ...lve.	■	■		Ψ	■	◈■	■	■	■		■	⊙	◈■	■	◈■
Broad Street	A M	A M	A M	A M	A M	A M	A M	A M	A M	†P M	P M	P M	P M	P M	P M
Broad St. Sub'n Sta.	*8 15	§8 45	†9 15	*9 45	†10 15	*10 45	*11 15	k11 30	*11 45	12 16	*12 45	*1 15	†1 45	*2 15	†2 45
Penna. Sta. (30th St.)	8 18	8 48	9 18	9 48	10 18	10 48	11 18	11 33	11 48	12 18	12 48	1 18	1 48	2 18	2 48
52d Street	8 23	8 53	9 24	9 53	10 23	10 53	11 23	– –	11 53	12 24	12 53	1 24	1 53	2 23	2 53
Overbrook +	8 26	8 56	9 27	9 56	10 26	10 56	11 26	– –	11 56	12 27	12 56	1 26	1 56	2 26	2 56
Merion +	8 28	8 58	9 29	9 58	10 28	10 58	11 28	– –	11 58	12 29	12 58	1 28	1 58	2 28	2 58
Narberth +	8 30	9 00	9 32	10 00	10 30	11 00	11 30	11 43	12 00	12 32	1 00	1 30	2 00	2 30	3 00
Wynnewood +	8 32	9 02	9 34	10 02	10 32	11 02	11 32	– –	12 02	12 34	1 02	1 32	2 02	2 32	3 02
Ardmore +	8 34	9 04	9 37	10 04	10 34	11 04	11 34	11 46	12 04	12 36	1 04	1 35	2 04	2 34	3 04
Haverford +	8 36	9 06	9 39	10 06	10 36	11 06	11 36	11 48	12 06	12 38	1 06	1 37	2 06	2 36	3 06
Bryn Mawr +	8 38	9 08	9 42	10 08	10 38	11 08	11 38	11 50	12 08	12 40	1 08	1 40	2 08	2 38	3 08
Rosemont +	8 40	9 10	9 44	10 10	10 40	11 10	11 40	– –	12 10	12 42	1 10	1 42	2 10	2 40	3 10
Villanova +	8 42	9 12	9 46	10 12	10 42	11 12	11 42	– –	12 12	12 44	1 12	1 44	2 12	2 42	3 12
Radnor +	8 44	9 14	9 48	10 14	10 44	11 14	11 44	– –	12 14	12 46	1 14	1 46	2 14	2 44	3 14
St. Davids +	8 46	9 16	9 51	10 16	10 46	11 16	11 46	11 54	12 16	12 48	1 16	1 48	2 16	2 46	3 16
Wayne +	8 48	9 18	9 53	10 18	10 48	11 18	11 48	11 56	12 18	12 50	1 18	1 50	2 18	2 48	3 18
Strafford +	8 50	9 20	9 56	10 20	10 50	11 20	11 50	– –	12 20	12 52	1 20	1 52	2 20	2 50	3 20
Devon +	8 52	9 22	9 59	10 22	10 52	11 22	11 52	– –	12 22	12 54	1 22	1 54	2 22	2 52	3 22
Berwyn +	8 54	9 24	10 02	10 24	10 54	11 24	11 54	12 01	12 24	12 56	1 24	1 56	2 24	2 54	3 24
Daylesford	– –	f9 26	– –	f10 26	f10 56	f11 26	f11 56	– –	f12 26	– –	f1 26	– –	f2 26	– –	f3 26
Paoli +	8 59	9 29	10 07	10 29	10 59	11 29	11 59	12 05	12 29	1 01	1 29	2 01	2 29	2 59	3 28
[ARRIVE	A M	A M	A M	A M	A M	A M	A M	P M	P M	P M	P M	P M	P M	P M	P M
								Bus.							
[LEAVE								☽							
Paoli								k12 10							
Green Tree								12 14							
Malvern +								12 17							
Frazer...(*P.R.R. Sta.*)								12 25							
Frazer (King & Goshen Rd.)								P M							

(*Eastern time.*)	Bus	5339	Bus	5341	5343	5345	5347	5349	5351	Bus
Philadelphia ......lve.				◈■	◈■	◈■	◈■	■	◈■	
Broad Street		P M		P M	P M	P M	P M	P M	P M	
Broad St. Sub'n Sta.		*3 15		‡3 43	†3 45	‡4 08	‡4 15	*4 20	‡4 30	
Penna. Sta. (30th St.)		3 18		3 46	3 48	4 11	4 18	4 23	4 33	
52d Street		3 24		– –	3 54	– –	– –	4 30	– –	
Overbrook		3 26		– –	3 56	– –	– –	4 32	– –	
Merion		3 28		– –	3 58	– –	– –	4 34	– –	
Narberth		3 30		– –	4 00	– –	4 28	4 36	– –	
Wynnewood	First trip September 8th.	3 33		– –	4 02	– –	4 30	4 38	– –	
Ardmore		3 35		3 58	4 04	4 24	4 32	4 40	4 45	
Haverford		3 37		4 00	4 06	4 26	4 34	4 42	4 47	
Bryn Mawr		3 39		4 02	4 08	4 28	4 36	4 44	4 49	
Rosemont		3 41		4 04	4 10	4 30	4 38	4 46	4 51	
Villanova		3 43		4 06	4 12	4 32	4 40	4 48	4 53	
Radnor		3 46		4 08	4 14	4 34	4 42	4 50	4 55	
St. Davids		3 48		4 10	4 16	4 36	4 44	4 52	4 57	
Wayne		3 50		4 12	4 18	4 38	4 46	4 53	4 59	
Strafford		3 52		4 14	4 20	4 40	4 48	4 55	5 01	
Devon		3 54		4 16	4 22	4 42	4 50	4 57	5 03	
Berwyn		3 56		4 18	4 24	4 44	4 53	4 59	5 05	
Daylesford		– –		f4 20	– –	f4 45	– –	f5 01	f5 07	
Paoli		4 01		4 23	4 29	4 49	4 57	5 04	5 10	
[ARRIVE		P M		P M	P M	P M	P M	P M	P M	
	☽		☽◈							☽◈
[LEAVE	P M		P M							P M
Paoli	‡3 32		‡4 20							†5 14
Green Tree	3 35		4 24							5 17
Malvern	3 38		4 27							5 33
Frazer....(*P.R.R. Sta.*)			4 36							
Frazer (King & Goshen Rd.)	3 44		P M							5 39
Immaculata	3 47									5 42
Morstein	3 49									5 44
Kirkland	3 52									5 47
Green Hill	3 54									5 49
Fern Hill	3 58									5 53
Biddle Street	4 02									5 57
West Chester	4 07									6 02
[ARRIVE	P M									P M

(*Eastern time.*)	5353	5355	5457	5357	5359	5361	5363	5365	5367	5369	5371	5373	5375	5377	5379
Philadelphia ......lve.	◈■	◈■	■	◈■	⊙	■	■	■	■	■	■	■	■	■	■
Broad Street	P M	P M	P M	P M	P M	P M	P M	P M	P M	P M	P M	P M	P M	P M	P M
Broad St. Sub'n Sta.	‡4 43	‡4 45	•4 45	‡5 00	*5 15	†5 45	*6 15	*6 45	*7 15	*7 45	*8 15	*8 50	*9 45	*10 25	*11 25
Penna. Sta. (30th St.)	4 46	4 48	4 48	5 03	5 18	5 48	6 18	6 48	7 18	7 48	8 18	8 53	9 48	10 28	11 28
52d Street	– –	4 54	4 54	– –	5 24	5 54	6 23	6 53	7 23	7 54	8 24	8 58	9 53	10 33	11 33
Overbrook	– –	4 56	4 56	– –	5 27	5 57	6 26	6 56	7 26	7 56	8 26	9 01	9 56	10 36	11 36
Merion	4 55	– –	4 58	– –	5 29	5 59	6 28	6 58	7 28	7 58	8 28	9 03	9 58	10 38	11 38
Narberth	– –	5 00	5 00	– –	5 31	6 01	6 30	7 00	7 30	8 00	8 30	9 05	10 00	10 40	11 40
Wynnewood	4 58	– –	5 02	– –	5 33	6 03	6 32	7 02	7 32	8 02	8 32	9 07	10 02	10 42	11 42
Ardmore	– –	5 03	5 04	5 15	5 36	6 05	6 34	7 04	7 34	8 04	8 34	9 09	10 04	10 44	11 44
Haverford	5 01	– –	5 06	5 17	5 37	6 07	6 36	7 06	7 36	8 06	8 36	9 11	10 06	10 46	11 46
Bryn Mawr	– –	5 06	5 08	5 19	5 39	6 09	6 38	7 08	7 38	8 08	8 38	9 13	10 08	10 48	11 48
Rosemont	5 04	– –	5 10	5 21	5 41	6 11	6 40	7 10	7 40	8 10	8 40	9 15	10 10	10 50	11 50
Villanova	– –	5 09	5 12	5 23	5 43	6 13	6 42	7 12	7 42	8 12	8 42	9 17	10 12	10 52	11 52
Radnor	5 08	– –	5 14	5 25	5 45	6 15	6 44	7 14	7 44	8 14	8 44	9 19	10 14	10 54	11 54
St. Davids	– –	5 12	5 16	5 27	5 47	6 17	6 46	7 16	7 46	8 16	8 46	9 21	10 16	10 56	11 56
Wayne	5 11	– –	5 18	5 29	5 49	6 19	6 48	7 18	7 48	8 18	8 48	9 23	10 18	10 58	11 58
Strafford	– –	5 15	5 20	5 31	5 51	6 21	6 50	7 20	7 50	8 20	8 50	9 25	10 20	11 00	12 00
Devon	5 15	5 17	5 22	5 33	5 53	6 23	6 52	7 22	7 52	8 22	8 52	9 27	10 22	11 02	12 02
Berwyn	5 17	5 19	5 24	5 35	5 55	6 25	6 54	7 24	7 54	8 24	8 54	9 29	10 24	11 04	12 04
Daylesford	– –	f5 21	f5 26	f5 38	f5 57	f6 27	– –	f7 26	f7 56	f8 26	– –	f9 31	f10 26	f11 06	f12 06
Paoli	5 22	5 24	5 29	5 41	6 00	6 30	6 59	7 29	7 59	8 30	8 59	9 34	10 28	11 09	12 09
[ARRIVE	P M	P M	P M	P M	P M	P M	P M	P M	P M	P M	P M	P M	P M	P M	A M
			Also runs September 7th.												

For Additional Trains between Philadelphia and West Chester, see Maryland Division, Table 76.

EXPLANATION OF SIGNS.

* Daily.
† Daily, except Sunday.
§ Sunday only.
f Stops only on signal or notice to agent or conductor to receive or discharge passengers.
k Saturday only.

‡ Daily, except Saturday and Sunday.
• Runs Saturday and Sunday only.
■ No baggage service.
◈ Will not run September 7th.
Ψ No baggage service week-days.

⊙ No baggage service Sunday.
☽ Bus on public highway; all classes of tickets good for passage over the Pennsylvania Railroad will be accepted on these busses.
+ Coupon stations.

For number of table upon which each station is located, see General Index of Stations in back part of Guide.

The Pennsylvania Railroad July 12, 1936 Table 97 schedule from the *August 1936 Official Guide of the Railways* shows local trains from Philadelphia to Paoli and five connecting bus trips from Paoli to West Chester plus two connecting bus trips from Paoli to Frazer. Buses, which replaced railroad passenger service between Paoli and West Chester, used the public highways, and Pennsylvania Railroad tickets were accepted on these buses.

286

PENNSYLVANIA RAILROAD

The Standard Railroad of the World

Table 97—PHILADELPHIA DIVISION.

Eastern Standard time. **Local Trains Paoli to Philadelphia—Bus Service West Chester to Paoli.** Eastern Region

July 12, 1936.	5302	5304	Bus	5306	Bus	5308	5310	5312	5314	5316	5318	Bus
[LEAVE			)◈									)◈
West Chester			†5 15									†6 33
Biddle Street			5 18									6 36
Fern Hill			5 22									6 40
Green Hill			5 26									6 44
Kirkland			5 28									6 46
Morstein			5 32									6 50
Immaculata			5 34		)◈							6 52
Frazer (King & Goshen Rd.)			5 37		A M							6 55
Frazer (*P. R.R. Sta.*)					†5 54							
Malvern			5 43		6 02							7 01
Green Tree			5 46		6 05							7 04
Paoli			5 50		6 09							7 08
[ARRIVE			A M		A M							A M
	◈⊙	⊙		⊙		⊡	◈	◈⊙	◈⊙	⊙	◈⊙	
[LEAVE	A M	A M		A M		A M	A M	A M	A M	A M	A M	
Paoli	†5 00	†5 16		*5 58		*6 20	†6 31	†6 42	†6 54	*7 00	†7 08	
Daylesford	f5 02	f5 18		6 02		f6 22	– –	f6 44	– –	f7 02	f7 10	
Berwyn	5 04	5 20		6 04		6 25	6 35	6 46	6 58	7 04	7 12	
Devon	5 06	5 22		6 06		6 28	6 37	6 48	7 00	7 06	7 14	
Strafford	f5 07	f5 24		6 08		6 30	6 39	6 50	7 02	7 08	7 16	
Wayne	5 09	5 26		6 10		6 32	6 41	6 52	7 04	7 10	7 18	
St. Davids	f5 11	5 28		6 12		6 34	6 43	6 54	7 06	7 12	7 20	
Radnor	f5 13	5 30		6 14		6 36	6 45	6 56	7 08	7 14	7 22	
Villanova	5 15	5 32		6 16		– –	6 47	6 58	7 10	7 16	7 24	
Rosemont	5 17	5 34		6 18		– –	6 49	7 00	7 12	7 18	7 26	
Bryn Mawr	5 19	5 36		6 20		6 40	6 51	7 02	7 14	7 20	7 28	
Haverford	5 21	5 38		6 22		6 42	6 53	7 04	7 16	7 22	7 30	
Ardmore	5 23	5 40		6 24		6 44	6 55	7 06	7 18	7 24	7 32	
Wynnewood	5 25	5 42		6 26		– –	6 57	7 08	– –	7 26	– –	
Narberth	5 27	5 44		6 28		6 46	6 59	7 10	7 21	7 28	– –	
Merion	5 29	5 46		6 30		– –	7 01	7 12	– –	7 30	– –	
Overbrook	5 31	5 48		6 32		– –	7 03	7 14	7 24	7 32	– –	
52d Street	5 33	5 50		6 35		– –	7 06	– –	– –	7 35	– –	
Pa. Sta. (30th St.)	5 39	5 57		6 42		6 57	7 12	7 22	7 32	7 42	7 45	
Broad St. Sub'n Sta.	5 42	6 00		6 45		7 00	7 15	7 25	7 35	7 45	7 48	
Broad Street	A M	A M		A M		A M	A M	A M	A M	A M	A M	
Philadelphia arr.												

July 12, 1936.	5508	5320	5322	5324	5326	5328	5330	5702	5332	5334	5336	5338	Bus.	5340
[LEAVE													)	
West Chester													k11 30	
Biddle Street													11 33	
Fern Hill													11 37	
Green Hill													11 41	
Kirkland													11 43	
Morstein													11 47	
Immaculata	◈												11 49	
Frazer (King & Goshen Rd.)	A M												11 52	
Frazer (*P. R.R. Sta.*)	†7 05													
Malvern	7 10												11 58	
Green Tree	7 14												12 01	
Paoli	7 17												12 05	
[ARRIVE	A M												P M	
	◈	◈⊙		◈⊙	⊙	⊙	◈⊙	⊙		◈⊙	♮	◈⊙		Ψ
[LEAVE	A M	A M	A M	A M	A M	A M	A M	A M	A M	A M	A M	A M		P M
Paoli	†7 17	†7 31	*8 04	†8 19	*8 49	*9 19	†9 49	§10 16	†10 16	†10 49	*11 19	†11 49		*12 22
Daylesford	– –	f7 33	f8 06	f8 21	– –	f9 21	f9 51	– –	– –	f10 51	f11 21	f11 51		f12 24
Berwyn	7 22	7 35	8 08	8 23	8 53	9 23	9 53	10 20	10 20	10 53	11 23	11 53		12 26
Devon	7 24	7 37	8 10	8 25	8 55	9 25	9 55	10 22	10 22	10 55	11 25	11 55		12 28
Strafford	– –	7 39	8 12	8 27	8 57	9 27	9 57	10 24	10 24	10 57	11 27	11 57		12 30
Wayne	7 28	7 41	8 14	8 29	8 59	9 29	9 59	10 26	10 26	10 59	11 29	11 59		12 32
St. Davids	– –	7 43	8 15	8 31	9 01	9 31	10 01	10 28	10 28	11 01	11 31	12 01		12 34
Radnor	– –	7 45	8 17	8 33	9 03	9 33	10 03	10 30	10 30	11 03	11 33	12 03		12 36
Villanova	7 32	7 47	8 19	8 35	9 05	9 35	10 05	10 32	10 32	11 05	11 35	12 05		12 38
Rosemont	– –	7 49	8 21	8 37	9 07	9 37	10 07	10 34	10 34	11 07	11 37	12 07		12 40
Bryn Mawr	7 36	7 51	8 23	8 39	9 09	9 39	10 09	10 36	10 37	11 09	11 39	12 09		12 42
Haverford	7 39	7 53	8 25	8 41	9 11	9 41	10 11	10 38	10 39	11 11	11 41	12 11		12 44
Ardmore	– –	7 55	8 27	8 43	9 13	9 43	10 13	10 40	10 42	11 13	11 43	12 13		12 46
Wynnewood	7 42	7 57	8 29	8 45	9 15	9 45	10 15	10 42	10 45	11 15	11 45	12 15		12 48
Narberth	– –	7 59	8 31	8 47	9 17	9 47	10 17	10 44	10 47	11 17	11 47	12 17		12 50
Merion	– –	8 01	8 33	8 49	9 19	9 49	10 19	10 46	10 49	11 19	11 49	12 19		12 52
Overbrook	7 46	8 03	8 35	8 51	9 21	9 51	10 21	10 48	10 51	11 21	11 51	12 21		12 54
52d Street	– –	8 06	8 38	8 54	– –	9 54	10 24	10 51	10 54	11 24	11 54	12 24		12 57
Pa. Sta. (30th St.)	7 55	8 12	8 44	9 00	9 30	10 00	10 30	10 57	11 00	11 30	12 00	12 30		1 02
Broad St. Sub'n Sta.	– –	8 15	8 47	9 03	9 33	10 03	10 33	11 00	11 03	11 33	12 03	12 33		1 05
Broad Street	7 59	A M	A M	A M	A M	A M	A M	A M	A M	A M	P M	P M		P M
Philadelphia arr.	A M													

(*Eastern time.*)	5342	5704	5344	5346	5348	5350	Bus	5352	5354	5356	Bus.
[LEAVE							)				)◈
West Chester							•2 40				†4 13
Biddle Street							2 43				4 16
Fern Hill							2 47				4 20
Green Hill							2 51				4 24
Kirkland							2 53				4 26
Morstein							2 57				4 30
Immaculata							2 59				4 32
Frazer (King & Goshen Rd.)							3 02				4 35
Frazer (*P. R.R. Sta.*)											
Malvern							3 08				4 41
Green Tree							3 11				4 44
Paoli							3 15				4 48
[ARRIVE							P M				P M
	◈⊙	⊙		◈⊙	⊙	◈⊙	**First trip September 8th.**	⊙	Ψ	♮	
[LEAVE	P M	P M	P M	P M	P M	P M		P M	P M	P M	
Paoli	†12 49	§1 15	†1 15	†1 49	*2 19	†2 49		*3 17	*3 47	*4 23	
Daylesford	f12 51	– –	f1 17	– –	f2 21	f2 51		f3 19	f3 49	– –	
Berwyn	12 53	1 19	1 19	1 53	2 23	2 53		3 21	3 51	4 27	
Devon	12 55	1 21	1 21	1 55	2 25	2 55		3 23	3 53	4 29	
Strafford	12 57	1 23	1 23	1 57	2 27	2 57		3 25	3 55	4 31	
Wayne	12 59	1 25	1 25	1 59	2 29	2 59		3 27	3 57	4 33	
St. Davids	1 01	1 27	1 27	2 01	2 31	3 01		3 29	3 59	4 35	
Radnor	1 03	1 29	1 29	2 03	2 33	3 03		3 31	4 01	4 37	
Villanova	1 05	1 31	1 31	2 05	2 35	3 05		3 33	4 03	4 39	
Rosemont	1 07	1 33	1 33	2 07	2 37	3 07		3 35	4 05	4 41	
Bryn Mawr	1 09	1 35	1 35	2 09	2 39	3 09		3 37	4 08	4 43	
Haverford	1 11	1 37	1 38	2 11	2 41	3 11		3 39	4 10	4 45	
Ardmore	1 13	1 39	1 41	2 13	2 43	3 13		3 41	4 12	4 47	
Wynnewood	1 15	1 41	1 44	2 15	2 45	3 15		3 43	4 15	4 49	
Narberth	1 17	1 43	1 47	2 17	2 47	3 17		3 45	4 17	4 51	
Merion	1 19	1 45	1 49	2 19	2 49	3 19		3 47	4 19	4 53	
Overbrook	1 21	1 47	1 51	2 21	2 51	3 21		3 49	4 21	4 55	
52d Street	d1 24	1 49	1 54	2 24	2 54	3 24		3 52	4 24	4 58	
Pa. Sta. (30th St.)	1 30	1 55	2 00	2 30	3 00	3 30		3 58	4 30	5 04	
Broad St. Sub'n Sta.	1 33	1 58	2 03	2 33	3 03	3 33		4 01	4 33	5 07	
Broad Street	P M	P M	P M	P M	P M	P M		P M	P M	P M	
Philadelphia arr.											

(*Eastern time.*)	5358	5360	5362	5364	5366	5368	5370	5372	5374	5376	5378			
[LEAVE														
West Chester														
Biddle Street														
Fern Hill														
Green Hill														
Kirkland														
Morstein														
Immaculata														
Frazer (King & Goshen Rd.)														
Frazer (*P. R.R. Sta.*)														
Malvern														
Green Tree														
Paoli														
[ARRIVE														
		⊙	⊙	◈Ψ	⊙	◈⊙	⊙	⊙	⊙	⊙	⊙			
[LEAVE	P M	P M	P M	P M	P M	P M	P M	P M	P M	P M	P M			
Paoli	†4 55	*5 31	*6 01	*6 31	*7 09	†7 31	*8 10	*9 01	*9 46	*10 31	*10 53			
Daylesford	– –	f5 33	f6 03	f6 33	– –	f7 33	f8 12	f9 03	– –	f10 33	f10 55			
Berwyn	4 59	5 35	6 05	6 35	7 13	7 35	8 14	9 05	9 50	10 35	10 57			
Devon	5 01	5 37	6 07	6 37	7 15	7 37	8 16	9 07	9 52	10 37	10 59			
Strafford	5 03	5 39	6 09	6 39	7 17	7 39	8 18	9 09	9 54	10 39	11 01			
Wayne	5 05	5 41	6 11	6 41	7 19	7 41	8 20	9 11	9 56	10 41	11 03			
St. Davids	5 07	5 43	6 13	6 43	7 20	7 43	8 22	9 13	9 58	10 43	11 04			
Radnor	5 09	5 46	6 15	6 45	7 22	7 45	8 24	9 15	10 00	10 45	11 06			
Villanova	5 11	5 47	6 17	6 47	7 24	7 47	8 26	9 17	10 02	10 47	11 08			
Rosemont	5 13	5 49	6 19	6 49	7 26	7 49	8 28	9 19	10 04	10 49	11 10			
Bryn Mawr	5 17	5 51	6 21	6 51	7 28	7 51	8 30	9 21	10 06	10 51	11 12			
Haverford	5 19	5 53	6 23	6 53	7 30	7 53	8 32	9 23	10 08	10 53	11 14			
Ardmore	5 22	5 55	6 25	6 55	7 32	7 55	8 34	9 25	10 10	10 55	11 16			
Wynnewood	5 25	5 57	6 27	6 57	7 34	7 57	8 35	9 27	10 12	10 57	11 18			
Narberth	5 27	5 59	6 29	6 59	7 36	7 59	8 38	9 29	10 14	10 59	11 20			
Merion	5 30	6 01	6 31	7 01	7 38	8 01	8 40	9 31	10 16	11 01	11 22			
Overbrook	5 32	6 03	6 33	7 03	7 40	8 03	8 42	9 33	10 18	11 03	11 24			
52d Street	5 35	6 06	6 36	7 06	7 42	8 06	8 44	9 36	10 21	11 06	11 26			
Pa. Sta. (30th St.)	5 42	6 12	6 42	7 12	7 47	8 12	8 49	9 42	10 27	11 12	11 32			
Broad St. Sub'n Sta.	5 45	6 15	6 45	7 15	7 50	8 15	8 52	9 45	10 30	11 15	11 35			
Broad Street	P M	P M	P M	P M	P M	P M	P M	P M	P M	P M	P M			
Philadelphia arr.														

For Additional Trains between West Chester and Philadelphia, see Maryland Division, Table 76.

Table 99—PHOENIXVILLE AND FRAZER. Bus Service.

	◈ Bus.	Mls	*July* 12, 1936.	Bus.	◈ Bus.
	†5 15 A M	0	lve.... + **Phoenixville** .....arr.	1 05 P M	5 10 P M
	5 25 "	2.7	Pickering	12 55 "	5 01 "
	5 29 "	4.4	Aldham	12 52 "	4 57 "
	5 33 "	5.5	+ Devault	12 47 "	4 53 "
	5 37 "	7.0	Sidley	12 43 "	4 49 "
	5 41 "	8.1	Bacton	12 39 "	4 46 "
	5 45 "	9.1	Swedesford Road	12 35 "	4 41 "
	5 50 A M	10.0	arr........ + **Frazer** ....lve.	k12 30 P M	•4 36 P M

All classes of tickets good over the Pennsylvania Railroad will be accepted on the above busses.

* Daily.
† Daily, except Sunday
§ Sunday only.
d Stops only to discharge passengers.
f Stops only on signal or notice to agent or conductor to receive or discharge passengers.
k Saturday only.
◈ Will not run September 7th.
♮ No baggage service weekdays.

• Daily, except Saturday and Sunday.
⊙ No baggage service.
Ψ No baggage service Sunday.
⊡ No baggage service except on holidays.
) Bus on public highway; all classes of tickets good for passage over the Pennsylvania Railroad will be accepted on these buses.
+ Coupon stations.

Five bus trips from West Chester to Paoli plus two bus trips from Frazer to Paoli made train connections for Philadelphia are shown on the Pennsylvania Railroad July 12, 1936 Table 97 schedule from the *August 1936 Official Guide of the Railways*. While railroad passenger service between West Chester and Paoli had been replaced by bus service, the Pennsylvania Railroad still had rail passenger service on the West Chester via Media to Philadelphia line.

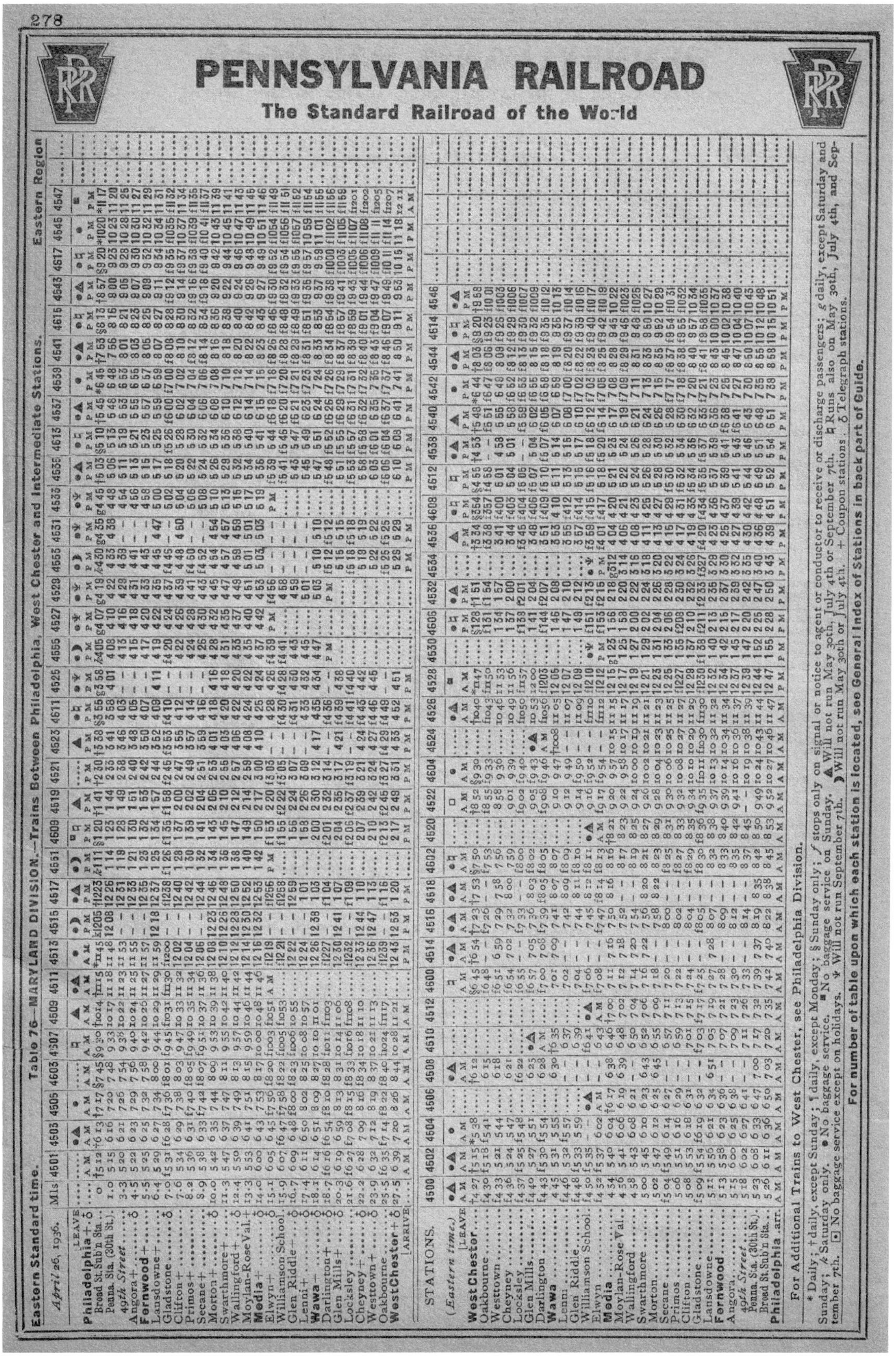

Pennsylvania Railroad April 26, 1936 Table 76 schedule from the *August 1936 Official Guide of the Railways* shows trains between Philadelphia and West Chester via Media. In spite of competition from the automobile and trolley lines, there was still excellent rail passenger service on this line.

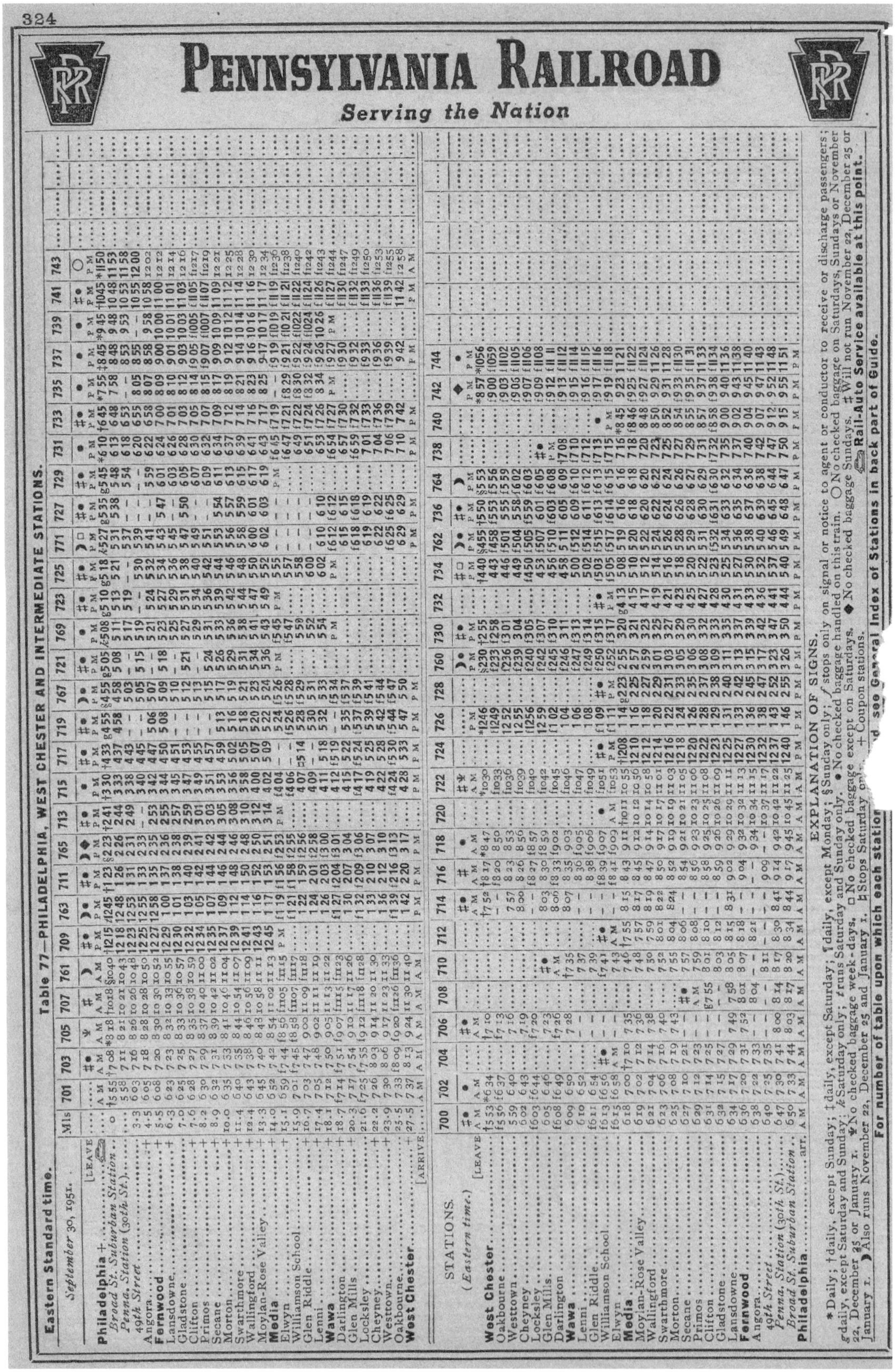

The September 30, 1951 Pennsylvania Railroad Table 77 schedule for the Philadelphia via Media to West Chester line from the *April 1952 Official Guide of the Railways* exhibited a decline from the April 26, 1936 schedule, but there was still an adequate amount of service during rush hours for commuters working in Philadelphia.

"SAFETY FIRST—DO NOT GET ON OR OFF MOVING TRAINS"

Philadelphia TO WEST CHESTER

		MONDAYS thru FRIDAYS, Except HOLIDAYS (see note below)									SATURDAYS; Feb. 17 (Washington's Birthday) March 28 (Good Friday)		
Miles		★ 0703	★ 0709	★ 0715	0723	0739	★ 0745	★ 0753	0759	0765	0705	0715	0743
		AM	AM	AM	AM	PM	PM	PM	PM	PM	AM	AM	PM
	Philadelphia												
.0	Lv Penn Center	6.00	7.55	9.25	11.25	3.25	4.54	5.34	6.30	7.55	7.00	9.25	4.25
------	" Penn Central Sta. (30th Street)	6.03	7.58	9.28	11.28	3.28	4.57	5.37	6.33	7.58	7.03	9.28	4.28
3.3	" Forty-ninth Street	6.07	8.02	------	------	3.32	------	------	------	------	7.07	------	4.33
4.5	" Angora	6.09	8.04	9.34	------	3.34	5.06	------	6.41	8.04	7.09	9.34	4.35
5.5	Lv Fernwood (Yeadon)	6.11	8.06	9.36	11.36	3.36	5.08	------	6.43	8.06	7.11	9.36	4.37
6.3	" Lansdowne	6.13	8.08	9.38	11.38	3.38	------	5.47	6.45	8.08	7.13	9.38	4.39
7.0	" Gladstone	------	------	------	11.39	3.39	------	------	6.47	8.09	------	------	4.40
7.6	" Clifton (Aldan)	6.16	8.12	9.41	11.41	3.41	------	------	6.49	8.11	7.16	9.41	4.42
8.2	Lv Primos	6.18	8.14	9.43	11.43	3.43	------	------	6.51	8.13	7.18	9.43	4.44
8.9	" Secane	6.20	8.16	9.45	11.45	3.45	5.14	5.54	6.53	8.15	7.20	9.45	4.46
10.0	" Morton (Rutledge)	6.22	8.18	9.47	11.47	3.47	5.17	5.57	6.55	8.17	7.22	9.47	4.48
11.4	" Swarthmore	6.24	8.20	9.49	11.49	3.49	5.20	6.00	6.57	8.19	7.24	9.49	4.50
12.4	Lv Wallingford	------	------	9.51	11.51	3.51	5.23	6.03	6.59	8.21	------	9.51	4.52
13.3	" Moylan—Rose Valley	------	•8.25	9.53	11.53	3.53	5.26	6.06	7.01	8.23	7.27	9.53	4.54
14.0	Ar **Media**	6.31	8.28	9.57	11.57	4.00	5.30	6.10	7.05	8.29	7.32	9.57	4.59
		Thru train	Change trains	Thru train	Thru train	Change trains	Change trains	Change trains	Thru train	Thru train	Thru train	Thru train	Thru train
14.0	Lv **Media**	6.31	8.29	9.57	11.57	4.02	5.32	6.13	7.05	8.29	7.32	9.57	4.59
15.1	" Elwyn	------	f 8.31	f 9.59	------	f 4.04	f 5.34	f 6.15	f 7.07	------	------	f 9.59	f 5.02
15.9	" Williamson School	------	------	f10.01	------	f 4.05	f 5.35	f 6.17	f 7.09	------	------	f10.01	------
16.7	" Glen Riddle	f 6.35	f 8.33	f10.03	f12.06	f 4.07	f 5.37	f 6.21	f 7.13	f 8.34	------	f10.03	f 5.05
17.4	Lv Lenni	------	f 8.35	f10.05	f12.08	f 4.09	f 5.39	f 6.23	f 7.15	------	f 7.39	f10.05	f 5.07
18.1	" Wawa	f 6.38	f 8.37	f10.07	f12.10	f 4.10	f 5.40	f 6.25	f 7.19	f 8.36	f 7.41	f10.07	f 5.08
18.7	" Darlington	------	------	f10.09	------	f 4.12	f 5.42	f 6.28	f 7.20	f 8.38	------	f10.09	f 5.10
20.3	Lv Glen Mills	f 6.42	f 8.41	f10.12	f12.14	f 4.16	f 5.46	f 6.32	f 7.24	f 8.42	f 7.45	f10.12	f 5.13
21.6	" Locksley	------	------	f10.15	------	f 4.19	f 5.49	f 6.36	------	------	------	f10.15	------
22.2	" Cheyney	f 6.46	f 8.46	f10.16	f12.19	f 4.21	f 5.51	f 6.38	f 7.28	f 8.48	f 7.50	f10.16	f 5.16
23.9	Lv Westtown	f 6.48	f 8.49	f10.18	f12.21	f 4.24	f 5.54	f 6.41	f 7.31	f 8.52	f 7.52	f10.18	f 5.18
27.0	" Nields Street	f 6.54	f 8.56	f10.23	f12.26	f 4.31	f 6.07	f 6.48	f 7.38	f 8.59	f 7.57	f10.23	f 5.23
27.5	Ar **West Chester**	6.58	8.59	10.25	12.28	4.34	6.10	6.51	7.41	9.02	8.00	10.25	5.28
		AM	AM	AM	PM	PM	PM	PM	PM	PM	AM	AM	PM

REFERENCE NOTES

HOLIDAYS—
New Year's Day, Memorial Day, Independence Day, Labor Day, Thanksgiving Day and Christmas Day, No Service.
Washington's Birthday and Good Friday. See Saturday Schedule.

★ Bargain one-way tickets are not honored on this train.
f Stops only on signal to receive or discharge passengers.

No service between Media and West Chester on Sundays and Holidays except Washington's Birthday and Good Friday. For complete service between Philadelphia and Media, see Form 37.

Responsibility—The Penn Central cannot assume responsibility for inconvenience, expense or damage resulting from errors in time tables, delayed trains, failure to make connections, or for shortage of equipment. The schedules shown in this time table are subject to change without notice.

For Further Information ask your ticket agent or phone—
Philadelphia, EV 7-6600—
6.00 A.M. to 11.45 P.M. Monday thru Fridays.
7.00 A.M. to 11.45 P.M. Saturday, Sunday and Holidays.
Media, 566-4070, 6.45 A.M. to 3.15 P.M. Mon. thru Fri.
West Chester, 696-0233, 6.45 A.M. to 3.15 P.M. Mon. thru Fri.
To Report lost articles, Philadelphia, 594-2131.
For SEPTA bus/rail service information, Philadelphia, DA 9-4800.

PRINTED IN U. S. A. 1st. ed. 50M 10-27-74 ALLEN, LANE & SCOTT, PHILA.

The Penn Central/SEPTA October 27, 1974 schedule from Philadelphia via Media to West Chester showed nine trains of which five were through trains on Monday through Friday and Saturday there were three through trains. There was no service on holidays except three trains on Washington's Birthday and Good Friday. (*Beth & Brian Keates collection*)

"SAFETY FIRST—DO NOT GET ON OR OFF MOVING

West Chester TO PHILADELPHIA

	MONDAYS thru FRIDAYS, Except HOLIDAYS (see note below)									SATURDAYS; Feb. 17 (Washington's Birthday) March 28 (Good Friday)		
	★ 0704	★ 0714	★ 0722	0730	0736	0744	0760	0768	0770	0726	0736	0786
	AM	AM	AM	AM	AM	PM	PM	PM	PM	AM	AM	PM
Lv **West Chester**	6.00	7.05	7.45	9.11	10.45	12.45	4.42	7.56	9.30	8.26	10.45	5.35
" Nields Street	f 6.01	f 7.06	f 7.46	f 9.12	f10.46	f12.46	f 4.43	f 7.57	f 9.31	f 8.27	f10.46	f 5.36
" Westtown	f 6.04	f 7.13	f 7.53	f 9.19	f10.51	f12.51	f 4.50	f 8.04	f 9.38	f 8.32	f10.51	f 5.41
Lv Cheyney	f 6.09	f 7.16	f 7.56	f 9.22	f10.54	f12.56	f 4.53	f 8.07	f 9.41	f 8.35	f10.54	f 5.44
" Locksley	f 6.10	f 7.18	f 7.58	------	f10.55	------	f 4.55	------	------	------	f10.55	------
" Glen Mills	f 6.14	f 7.21	f 8.01	f 9.26	f10.58	f 1.01	f 4.58	f 8.11	f 9.45	f 8.38	f10.58	f 5.47
Lv Darlington	------	f 7.25	f 8.05	f 9.30	f11.02	------	f 5.02	------	------	f 8.40	f11.02	------
" Wawa	f 6.16	f 7.26	f 8.06	f 9.31	f11.03	f 1.04	f 5.03	f 8.14	f 9.47	f 8.42	f11.03	f 5.50
" Lenni	f 6.18	f 7.28	f 8.08	f 9.33	f11.04	f 1.06	f 5.05	------	------	f 8.44	f11.04	f 5.51
Lv Glen Riddle	f 6.20	f 7.30	f 8.10	f 9.35	f11.06	f 1.07	f 5.07	------	f 9.49	f 8.46	f11.06	f 5.52
" Williamson School	------	f 7.32	f 8.12	f 9.37	f11.08	------	f 5.09	------	------	------	f11.08	f 5.54
" Elwyn	------	f 7.33	f 8.13	------	f11.10	------	f 5.10	------	f 9.55	------	f11.10	------
Ar **Media**	6.30	7.38	8.17	9.41	11.15	1.15	5.14	8.28	10.00	8.55	11.15	6.05
	Thru train	Thru train	Change trains	Thru train	Thru train	Thru train	Change trains	Thru train	Thru train	Thru train	Thru train	Thru train
Lv **Media**	6.30	7.38	8.18	9.45	11.15	1.15	5.15	8.28	10.00	8.55	11.15	6.05
" Moylan—Rose Valley	6.32	↓	8.20	9.47	11.17	1.17	5.17	8.30	10.02	8.57	11.17	6.07
" Wallingford	6.34	↓	8.23	9.49	11.19	1.19	5.19	8.32	10.04	8.59	11.19	6.09
Lv Swarthmore	6.37	↓	8.25	9.51	11.21	1.21	5.21	8.34	10.06	9.01	11.21	6.11
" Morton (Rutledge)	6.39	↓	8.27	9.53	11.23	1.23	5.23	8.36	10.08	9.03	11.23	6.13
" Secane	6.41	↓	8.30	9.55	11.25	1.25	5.25	8.38	10.10	9.05	11.25	6.15
" Primos	6.43	↓	------	9.57	11.27	1.27	5.27	8.40	10.12	9.07	11.27	6.17
Lv Clifton (Aldan)	6.45	↓	------	9.59	11.29	1.29	5.29	8.42	10.14	9.09	11.29	6.19
" Gladstone	6.47	↓	------	10.00	11.30	1.30	5.30	8.43	10.15	9.11	11.30	------
" Lansdowne	6.49	↓	8.35	10.02	11.32	1.32	5.32	8.45	10.17	9.13	11.32	6.22
" Fernwood (Yeadon)	6.51	↓	------	10.04	11.34	1.34	5.34	8.47	10.19	9.15	11.34	6.24
Lv Angora	6.53	↓	------	10.06	------	------	5.36	8.49	10.21	9.17	------	6.26
" Forty-ninth Street	6.55	↓	------	10.08	------	------	5.38	------	------	------	------	------
Philadelphia												
Ar Penn Central Sta. (30th Street)	7.02	8.02	8.46	10.13	11.43	1.43	5.46	8.56	10.28	9.26	11.43	6.33
Ar Penn Center	7.06	8.06	8.50	10.17	11.47	1.47	5.50	9.00	10.32	9.30	11.47	6.37
	AM	AM	AM	AM	AM	PM	PM	PM	PM	AM	AM	PM

REFERENCE NOTES

HOLIDAYS—

New Year's Day, Memorial Day, Independence Day, Labor Day, Thanksgiving Day and Christmas Day, No Service.

Washington's Birthday and Good Friday. See Saturday Schedule.

★ Bargain one-way tickets are not honored on this train.

f Stops only on signal to receive or discharge passengers.

No service between Media and West Chester on Sundays and Holidays except Washington's Birthday and Good Friday. For complete service between Philadelphia and Media, see Form 37.

J. K. SHOEMAKER
Corridor Manager-Philadelphia District
Northeast Corridor Region
Penn Central Station (30th Street)
Philadelphia, Pa. 19104

The Penn Central/SEPTA October 27, 1974 schedule from West Chester via Media to Philadelphia showed nine trains of which seven were through trains on Monday through Friday and Saturday there were three through trains. There was no service on holidays except three trains on Washington's Birthday and Good Friday. (*Beth & Brian Keates collection*)

West Chester, Elwyn and Media TO PHILADELPHIA **MONDAYS Thru FRIDAYS Except HOLIDAYS**

Train Number		WEST CHESTER	NIELDS STREET	WESTTOWN	CHEYNEY	LOCKSLEY	GLEN MILLS	DARLINGTON	WAWA	LENNI	GLEN RIDDLE	WILLIAMSON SCHOOL	ELWYN	MEDIA	MOYLAN-ROSE VALLEY	WALLINGFORD	SWARTHMORE	MORTON	SECANE	PRIMOS	CLIFTON	GLADSTONE	LANSDOWNE	FERNWOOD	ANGORA	FORTY-NINTH STREET	30th ST. STATION	PENN CENTER	
702	AM	...	...	...	...	...	...	...	...	...	...	...	...	5.55	5.56	5.59	6.01	6.03	6.05	6.07	6.09	6.11	6.13	6.15	6.17	6.19	6.26	6.30	AM
0704	AM	5.57	f 5.59	f 6.04	f 6.07	...	f 6.12	...	f 6.15	f 6.17	f 6.19	...	6.23	6.27	6.28	6.31	6.33	6.35	6.37	6.39	6.41	6.43	6.45	6.47	6.49	6.51	7.00	7.04	AM
1706	AM	...	...	...	...	...	...	...	...	...	...	...	6.53	6.56	6.57	7.00	7.02	7.04	7.07	7.09	7.11	7.12	7.14	7.17	7.19	7.21	7.31	7.35	AM
708	AM	...	...	...	...	...	...	...	...	...	...	...	...	7.10	7.11	7.14	7.17	7.19	7.22	...	...	...	...	...	...	...	7.36	7.40	AM
1710	AM	...	...	...	...	...	...	...	...	...	...	...	7.14	7.17	...	...	...	...	7.25	7.27	7.29	7.30	7.32	7.35	7.37	7.39	7.49	7.53	AM
712	AM	...	...	...	...	...	...	...	...	...	...	...	...	7.27	7.28	7.31	7.34	7.37	7.40	...	...	...	...	...	...	...	7.56	8.00	AM
0714	AM	7.07	f 7.08	f 7.13	f 7.16	f 7.18	f 7.21	f 7.24	f 7.25	f 7.28	f 7.30	f 7.32	7.34	7.38	...	...	...	...	...	...	...	...	...	...	...	...	8.02	8.06	AM
716	AM	...	...	...	...	...	...	...	...	...	...	...	...	7.41	...	...	...	...	7.52	7.54	7.56	7.58	8.00	8.03	8.05	8.07	8.14	8.18	AM
1718	AM	...	...	...	...	...	...	...	...	...	...	...	7.50	7.53	7.54	7.57	7.59	8.02	8.04	...	...	...	...	...	...	...	8.23	8.27	AM
2720	AM	...	...	...	...	...	...	...	...	...	...	...	...	...	...	...	...	...	8.12	8.14	8.16	8.18	8.20	8.22	8.24	8.26	8.33	8.37	AM
0722	AM	7.40	f 7.41	f 7.46	f 7.49	f 7.50	f 7.53	...	f 7.58	f 8.00	f 8.02	f 8.04	8.06	M 8.12	8.13	8.16	8.18	8.20	8.23	...	...	...	8.29	...	...	...	8.40	8.44	AM
724	AM	...	...	...	...	...	...	...	...	...	...	...	...	8.38	8.39	8.42	8.45	8.47	8.50	8.52	8.54	8.56	8.58	9.00	9.02	9.04	9.11	9.15	AM
1726	AM	...	...	...	...	...	...	...	...	...	...	...	8.53	8.55	8.56	8.59	9.01	9.03	9.05	9.07	9.09	9.11	9.13	9.15	...	...	9.26	9.30	AM
728	AM	...	...	...	...	...	...	...	...	...	...	...	...	9.20	9.21	9.24	9.26	9.28	9.30	9.32	9.34	9.36	9.38	9.40	...	...	9.50	9.54	AM
0730	AM	9.14	f 9.15	f 9.20	f 9.24	...	f 9.27	f 9.29	f 9.30	f 9.32	f 9.34	f 9.36	9.40	9.45	9.46	9.49	9.51	9.53	9.55	9.57	9.59	10.00	10.02	10.04	10.06	10.08	10.15	10.19	AM
732	AM	...	...	...	...	...	...	...	...	...	...	...	...	10.15	10.16	10.19	10.21	10.23	10.25	10.27	10.29	10.30	10.32	10.34	...	...	10.45	10.49	AM
1734	AM	...	...	...	...	...	...	...	...	...	...	...	10.40	10.45	10.46	10.49	10.51	10.53	10.55	10.57	10.59	11.00	11.02	11.04	11.06	11.08	11.15	11.19	AM
0736	AM	10.45	f10.46	f10.51	f10.54	...	f10.58	...	f11.03	f11.04	f11.06	f11.08	11.10	11.15	11.16	11.19	11.21	11.23	11.25	11.27	11.29	11.30	11.32	11.34	...	...	11.45	11.49	AM
1738	AM	...	...	...	...	...	...	...	...	...	...	...	11.40	11.45	11.46	11.49	11.51	11.53	11.55	11.57	11.59	12.00	12.02	12.04	12.06	12.08	12.15	12.19	PM
740	PM	...	...	...	...	...	...	...	...	...	...	...	...	12.15	12.16	12.19	12.21	12.23	12.25	12.27	12.29	12.30	12.32	12.34	...	...	12.45	12.49	PM
742	PM	...	...	...	...	...	...	...	...	...	...	...	...	12.45	12.46	12.49	12.51	12.53	12.55	12.57	12.59	1.00	1.02	1.04	1.06	1.08	1.15	1.19	PM
0744	PM	12.45	f12.46	f12.50	f12.53	...	f12.56	...	f12.59	f 1.01	f 1.03	...	1.07	1.15	1.16	1.19	1.21	1.23	1.25	1.27	1.29	1.30	1.32	1.34	...	...	1.45	1.49	PM
1746	PM	...	...	...	...	...	...	...	...	...	...	...	1.40	1.45	1.46	1.49	1.51	1.53	1.55	1.57	1.59	2.00	2.02	2.04	2.06	2.08	2.15	2.19	PM
748	PM	...	...	...	...	...	...	...	...	...	...	...	...	2.15	2.16	2.19	2.21	2.23	2.25	2.27	2.29	2.30	2.32	2.34	...	...	2.45	2.49	PM
1750	PM	...	...	...	...	...	...	...	...	...	...	...	2.40	2.45	2.46	2.49	2.51	2.53	2.55	2.57	2.59	3.00	3.02	3.04	3.06	3.08	3.15	3.19	PM
0752	PM	2.45	f 2.46	f 2.51	f 2.55	...	f 2.58	...	f 3.00	f 3.02	f 3.04	f 3.06	3.10	3.15	3.16	3.19	3.21	3.23	3.25	3.27	3.29	3.30	3.32	3.34	3.36	3.38	3.45	3.49	PM
1754	PM	...	...	...	...	...	...	...	...	...	...	...	3.45	3.50	3.51	3.54	3.56	3.58	4.00	4.02	4.04	4.05	4.07	4.09	4.11	4.13	4.20	4.24	PM
756	PM	...	...	...	...	...	...	...	...	...	...	...	...	4.15	4.16	4.19	4.21	4.23	4.25	4.27	4.29	4.30	4.32	4.34	4.36	4.38	4.46	4.50	PM
1758	PM	...	...	...	...	...	...	...	...	...	...	...	4.45	4.48	4.49	4.52	4.54	4.57	5.00	5.02	5.04	5.05	5.07	5.09	5.11	5.13	5.22	5.26	PM
0760	PM	4.40	f 4.41	f 4.46	f 4.49	...	f 4.53	f 4.56	f 4.58	f 5.00	f 5.02	f 5.04	5.06	M 5.12	5.13	5.16	5.18	5.21	5.24	5.26	5.28	5.30	5.31	5.33	5.35	5.37	5.44	5.48	PM
1762	PM	...	...	...	...	...	...	...	...	...	...	...	5.42	5.45	5.46	5.49	5.51	5.53	5.55	5.57	5.59	...	6.02	6.04	6.06	6.08	6.15	6.19	PM
764	PM	...	...	...	...	...	...	...	...	...	...	...	...	6.20	6.21	6.24	6.26	6.28	6.30	6.32	6.34	6.36	6.38	6.40	6.42	...	6.49	6.53	PM
0766	PM	6.50	f 6.51	f 6.56	f 7.00	...	f 7.03	f 7.06	f 7.08	f 7.10	f 7.12	...	7.15	7.20	7.21	7.24	7.26	7.28	7.30	7.32	7.34	7.35	7.37	7.39	7.41	7.43	7.50	7.54	PM
0768	PM	8.05	f 8.06	f 8.11	f 8.15	...	f 8.18	...	f 8.20	...	...	...	8.32	8.35	8.36	8.39	8.41	8.43	8.45	8.47	8.49	8.51	8.53	8.55	8.57	...	9.05	9.09	PM
1770	PM	...	...	...	...	...	...	...	...	...	...	...	8.55	9.00	9.01	9.04	9.06	9.08	9.10	9.12	9.14	...	9.17	9.19	9.21	...	9.30	9.34	PM
0776	PM	9.30	f 9.31	f 9.38	f 9.41	...	f 9.45	...	f 9.47	...	f 9.49	...	9.55	10.00	10.01	10.04	10.06	10.08	10.10	10.12	10.14	10.15	10.17	10.19	10.21	...	10.30	10.34	PM
1778	PM	...	...	...	...	...	...	...	...	...	...	...	11.00	11.05	11.06	11.09	11.11	11.13	11.15	11.17	11.19	11.20	11.22	11.24	11.26	...	11.35	11.39	PM
780	PM	...	...	...	...	...	...	...	...	...	...	...	...	11.55	11.56	11.59	12.01	12.03	12.05	12.07	12.09	...	12.12	12.14	12.16	...	12.25	12.29	AM

■ Bargain tickets not accepted on these trains.

The April 27, 1980 Monday through Friday West Chester via Media to Philadelphia (Operated by Conrail under contract with SEPTA) commuter rail service showed eleven trips from West Chester via Media to Philadelphia, twenty-four trips from Elwyn to Philadelphia, and thirty-seven trips from Media to Philadelphia. Two of the trips from West Chester required changing trains at Media to continue to Philadelphia. (*Jose Escamilla collection*)

Philadelphia TO MEDIA, ELWYN and WEST CHESTER   MONDAYS Thru FRIDAYS Except HOLIDAYS

Train Number		PENN CENTER	30th ST. STATION	FORTY-NINTH STREET	ANGORA	FERNWOOD	LANSDOWNE	GLADSTONE	CLIFTON	PRIMOS	SECANE	MORTON	SWARTHMORE	WALLINGFORD	MOYLAN-ROSE VALLEY	MEDIA	ELWYN	WILLIAMSON SCHOOL	GLEN RIDDLE	LENNI	WAWA	DARLINGTON	GLEN MILLS	LOCKSLEY	CHEYNEY	WESTTOWN	NIELDS STREET	WEST CHESTER	
0703	AM	6.00	6.03	6.07	6.09	6.11	6.13	6.14	6.16	6.18	6.20	6.22	6.24		6.26	6.31	6.35		f 6.37		f 6.41		f 6.44		f 6.48	f 6.50	f 6.55	6.58	AM
1705	AM	6.50	6.53	6.57	6.59	7.01	7.03		7.06	7.08	7.10	7.12	7.14		7.17	7.21	7.25												AM
707	AM	7.25	7.28	7.32	7.34	7.36	7.38	7.39	7.41	7.43	7.45	7.47	7.49	7.51	7.53	7.59													AM
0709	AM	7.55	7.58	8.02			8.07				8.12	8.15	8.17	8.19	8.21	M 8.26	8.30		f 8.32	f 8.34	f 8.36		f 8.40		f 8.45	f 8.47	f 8.52	8.55	AM
1711	AM	8.05	8.08		8.14	8.16	8.18	8.19	8.21	8.23	8.25					8.40	8.44												AM
713	AM	8.28	8.31	8.35	8.37	8.39	8.41		8.44	8.46	8.48	8.50	8.52	8.54	8.56	9.03													AM
715	AM	8.55	8.58	9.02	9.04	9.06	9.08	9.09	9.11	9.13	9.15	9.17	9.19	9.21	9.23	9.29													AM
0715	AM	9.25	9.28		9.34	9.36	9.38		9.41	9.43	9.45	9.47	9.49	9.51	9.53	9.57	10.00	f10.02	f10.04	f10.06	f10.08	f10.10	f10.12	f10.14	f10.16	f10.18	f10.22	10.27	AM
1717	AM	9.55	9.58	10.02	10.04	10.06	10.08	10.09	10.11	10.13	10.15	10.17	10.19	10.21	10.23	10.27	10.32												AM
719	AM	10.25	10.28			10.36	10.38	10.39	10.41	10.43	10.45	10.47	10.49	10.51	10.53	10.59													AM
1721	AM	10.55	10.58	11.02	11.04	11.06	11.08	11.09	11.11	11.13	11.15	11.17	11.19	11.21	11.23	11.27	11.32												AM
0723	AM	11.25	11.28			11.36	11.38	11.39	11.41	11.43	11.45	11.47	11.49	11.51	11.53	11.57	12.00	f12.02	f12.04	f12.06	f12.08	f12.10	f12.12		f12.16	f12.18	f12.22	12.27	PM
725	AM	11.55	11.58	12.02	12.04	12.06	12.08	12.09	12.11	12.13	12.15	12.17	12.19	12.21	12.23	12.29													PM
727	PM	12.25	12.28			12.36	12.38	12.39	12.41	12.43	12.45	12.47	12.49	12.51	12.53	12.59													PM
1729	PM	12.55	12.58	1.02	1.04	1.06	1.08	1.09	1.11	1.13	1.15	1.17	1.19	1.21	1.23	1.27	1.32												PM
0731	PM	1.25	1.28			1.36	1.38	1.39	1.41	1.43	1.45	1.47	1.49	1.51	1.53	1.57	2.00	f 2.02	f 2.04	f 2.06	f 2.08	f 2.10	f 2.12	f 2.14	f 2.16	f 2.18	f 2.22	2.27	PM
1733	PM	1.55	1.58	2.02	2.04	2.06	2.08	2.09	2.11	2.13	2.15	2.17	2.19	2.21	2.23	2.27	2.32												PM
735	PM	2.25	2.28			2.36	2.38	2.39	2.41	2.43	2.45	2.47	2.49	2.51	2.53	3.00													PM
1737	PM	2.55	2.58	3.02	3.04	3.06	3.08	3.09	3.11	3.13	3.15	3.17	3.19	3.21	3.23	3.27	3.32												PM
0739	PM	3.25	3.28	3.32	3.34	3.36	3.38	3.39	3.41	3.43	3.45	3.47	3.49	3.51	3.53	4.00	4.03	f 4.05	f 4.07	f 4.09	f 4.10	f 4.12	f 4.15	f 4.17	f 4.19	f 4.22	f 4.27	4.33	PM
1741	PM	3.55	3.58	4.03	4.05	4.07	4.10	4.13	4.15	4.18	4.20	4.23	4.25	4.27	4.29	4.31	4.35												PM
743	PM	4.15	4.19								4.33	4.36	4.39	4.42	4.44	4.50													PM
2745	PM	4.20	4.24	4.29	4.31	4.33	4.35	4.37	4.39	4.41	4.45																		PM
1777	PM	4.40	4.44				4.52	4.54	4.56	4.59	5.01	5.04	5.07	5.10	5.13	5.17	5.22												PM
0745	PM	4.55	4.59		5.06	5.08					5.14	5.17	5.20	5.23	5.26	M 5.30	5.35	f 5.36	f 5.37	f 5.39	f 5.40	f 5.42	f 5.46	f 5.49	f 5.51	f 5.54	f 6.01	6.04	PM
747	PM	5.03	5.07			5.15	5.17	5.20	5.23	5.26	5.29					5.42													PM
1749	PM	5.15	5.19								5.35	5.40	5.43	5.46	5.49	5.54	6.00												PM
2751	PM	5.20	5.24	5.29	5.30	5.32	5.35	5.38	5.40	5.42	5.44																		PM
0753	PM	5.35	5.39				5.47				5.54	5.57	6.00	6.03	6.06	6.10	6.14	f 6.15	f 6.17	f 6.19	f 6.21	f 6.22	f 6.26	f 6.29	f 6.31	f 6.34	f 6.39	6.42	PM
755	PM	5.45	5.49	5.53	5.55	5.57	6.00	6.02	6.04	6.06	6.09	6.11	6.13	6.15	6.17	6.24													PM
1757	PM	6.05	6.08	6.13	6.15	6.17	6.19	6.21	6.23	6.25	6.27	6.30	6.32	6.34	6.36	6.39	6.44												PM
1759	PM	6.30	6.33		6.39	6.41	6.43	6.45	6.47	6.49	6.51	6.53	6.55	6.57	6.59	7.05	7.10												PM
0761	PM	6.50	6.53		7.01	7.03	7.05	7.07	7.09	7.11	7.13	7.15	7.17	7.19	7.21	7.25	7.29	f 7.31	f 7.33	f 7.35	f 7.37	f 7.38	f 7.40		f 7.44	f 7.47	f 7.52	f 7.55	PM
1763	PM	7.25	7.28	7.32	7.34	7.36	7.38	7.39	7.41	7.43	7.45	7.47	7.49	7.51	7.53	8.00	8.03												PM
0765	PM	8.05	8.08		8.14	8.16	8.18	8.20	8.22	8.24	8.26	8.28	8.30	8.32	8.34	8.38	8.41		f 8.44		f 8.46	f 8.48	f 8.52		f 8.56	f 9.00	f 9.05	9.09	PM
767	PM	9.00	9.03		9.09	9.11	9.13	9.14	9.16	9.18	9.20	9.22	9.24	9.26	9.28	9.34													PM
1769	PM	10.10	10.13		10.19	10.21	10.23	10.24	10.26	10.28	10.30	10.32	10.34	10.36	10.38	10.45	10.49												PM
771	PM	11.00	11.03	11.07	11.09	11.11	11.13	11.14	11.16	11.18	11.20	11.22	11.24	11.26	11.28	11.35													PM
773	PM	11.55	11.58		12.04	12.06	12.08	12.09	12.11	12.13	12.15	12.17	12.19	12.21	12.23	12.30													AM
701	AM	12.55	12.58		1.04	1.06	1.08	1.09	1.11	1.13	1.15	1.17	1.19	1.21	1.23	1.30													AM

Bargain tickets not accepted on these trains.

The April 27, 1980 Monday through Friday Philadelphia via Media to West Chester (Operated by Conrail under contract with SEPTA) commuter rail service showed ten trips from Philadelphia via Media to West Chester, twenty-four trips from Philadelphia to Elwyn, and thirty-eight trips from Philadelphia to Media. Two of the trips from West Chester required changing trains at Media to continue to Philadelphia. (*Jose Escamilla collection*)

West Chester, Elwyn and Media TO PHILADELPHIA — SATURDAYS and Washington's Birthday and Good Friday

Train Number		West Chester	Nields Street	Westtown	Cheyney	Locksley	Glen Mills	Darlington	Wawa	Lenni	Glen Riddle	Williamson School	Elwyn	Media	Moylan-Rose Valley	Wallingford	Swarthmore	Morton	Secane	Primos	Clifton	Gladstone	Lansdowne	Fernwood	Angora	Forty-Ninth Street	30th St. Station	Penn Center	
702	AM													5.55	5.56	5.59	6.01	6.03	6.05	6.07	6.09	6.11	6.13	6.15	6.17	6.19	6.26	6.30	AM
772	AM													6.45	6.46	6.49	6.51	6.53	6.55	6.57	6.59	7.01	7.03	7.05	7.07		7.16	7.20	AM
1774	AM												7.11	7.14	7.15	7.18	7.20	7.22	7.24	7.26	7.28	7.30	7.32	7.34	7.36		7.44	7.48	AM
716	AM													7.41	7.42	7.45	7.47	7.49	7.52	7.54	7.56	7.58	8.00	8.03	8.05	8.07	8.14	8.18	AM
1782	AM												8.14	8.19	8.20	8.23	8.25	8.27	8.29	8.31	8.33	8.35	8.37	8.39	8.41		8.51	8.55	AM
0726	AM	8.25	f 8.26	f 8.31	f 8.35		f 8.38	f 8.40	f 8.42	f 8.44	f 8.46		8.50	8.55	8.56	8.59	9.01	9.03	9.05	9.07	9.09	9.11	9.13	9.15	9.17		9.26	9.30	AM
728	AM													9.20	9.21	9.24	9.26	9.28	9.30	9.32	9.34	9.36	9.38	9.40			9.50	9.54	AM
1730	AM												9.40	9.45	9.46	9.49	9.51	9.53	9.55	9.57	9.59	10.00	10.02	10.04	10.06	10.08	10.15	10.19	AM
732	AM													10.15	10.16	10.19	10.21	10.23	10.25	10.27	10.29	10.30	10.32	10.34			10.45	10.49	AM
734	AM													10.45	10.46	10.49	10.51	10.53	10.55	10.57	10.59	11.00	11.02	11.04	11.06		11.15	11.19	AM
0736	AM	10.45	f10.46	f10.51	f10.54	f10.55	f10.58	f11.02	f11.03	f11.04	f11.06	f11.08	11.10	11.15	11.16	11.19	11.21	11.23	11.25	11.27	11.29	11.30	11.32	11.34			11.45	11.49	AM
738	AM													11.45	11.46	11.49	11.51	11.53	11.55	11.57	11.59	12.00	12.02	12.04	12.06	12.08	12.15	12.19	PM
1740	PM												12.10	12.15	12.16	12.19	12.21	12.23	12.25	12.27	12.29	12.30	12.32	12.34			12.45	12.49	PM
742	PM													12.45	12.46	12.49	12.51	12.53	12.55	12.57	12.59	1.00	1.02	1.04	1.06	1.08	1.15	1.19	PM
1744	PM												1.10	1.15	1.16	1.19	1.21	1.23	1.25	1.27	1.29	1.30	1.32	1.34	1.36		1.45	1.49	PM
746	PM													1.45	1.46	1.49	1.51	1.53	1.55	1.57	1.59	2.00	2.02	2.04	2.06	2.08	2.15	2.19	PM
1748	PM												2.10	2.15	2.16	2.19	2.21	2.23	2.25	2.27	2.29	2.30	2.32	2.34			2.45	2.49	PM
750	PM													2.45	2.46	2.49	2.51	2.53	2.55	2.57	2.59	3.00	3.02	3.04	3.06	3.08	3.15	3.19	PM
1754	PM												3.45	3.50	3.51	3.54	3.56	3.58	4.00	4.02	4.04	4.05	4.07	4.09	4.11	4.13	4.20	4.24	PM
1758	PM												4.45	4.48	4.49	4.52	4.54	4.57	5.00	5.02	5.04	5.05	5.07	5.09	5.11	5.13	5.22	5.26	PM
760	PM													5.12	5.13	5.16	5.18	5.21	5.24	5.26	5.28	5.30	5.31	5.33	5.35	5.37	5.44	5.48	PM
0786	PM	5.35	f 5.36	f 5.41	f 5.45		f 5.48		f 5.50	f 5.52	f 5.54	f 5.56	6.00	6.05	6.06	6.09	6.11	6.13	6.15	6.17	6.19		6.22	6.24	6.26		6.35	6.39	PM
1766	PM												6.55	7.00	7.01	7.04	7.06	7.08	7.10	7.12	7.14	7.15	7.17	7.19	7.21	7.23	7.30	7.34	PM
768	PM													8.00	8.01	8.04	8.06	8.08	8.10	8.12	8.14	8.15	8.17	8.19	8.21		8.30	8.34	PM
770	PM													9.00	9.01	9.04	9.06	9.08	9.10	9.12	9.14		9.17	9.19	9.21		9.30	9.34	PM
1776	PM												9.55	10.00	10.01	10.04	10.06	10.08	10.10	10.12	10.14	10.15	10.17	10.19	10.21	10.23	10.30	10.34	PM
700	AM													12.10	12.11	12.14	12.16	12.18	12.20	12.22	12.24		12.27	12.29	12.31	12.33	12.40	12.44	AM

SUNDAYS and New Year's, Memorial, Independence, Labor, Thanksgiving and Christmas Days

NO SERVICE BETWEEN WEST CHESTER-ELWYN AND MEDIA ON SUNDAYS AND HOLIDAYS LISTED.

Train Number		West Chester	Nields Street	Westtown	Cheyney	Locksley	Glen Mills	Darlington	Wawa	Lenni	Glen Riddle	Williamson School	Elwyn	Media	Moylan-Rose Valley	Wallingford	Swarthmore	Morton	Secane	Primos	Clifton	Gladstone	Lansdowne	Fernwood	Angora	Forty-Ninth Street	30th St. Station	Penn Center	
774	AM													7.10	7.11	7.14	7.16	7.18	7.20	7.22	7.24	7.26	7.28	7.30	7.32		7.39	7.43	AM
726	AM													8.55	8.56	8.59	9.01	9.03	9.05	9.07	9.09	9.11	9.13	9.15	9.17		9.26	9.30	AM
734◈	AM												◈	10.45	10.46	10.49	10.51	10.53	10.55	10.57	10.59	11.00	11.02	11.04	11.06		11.15	11.19	AM
738	AM													11.45	11.46	11.49	11.51	11.53	11.55	11.57	11.59	12.00	12.02	12.04	12.06	12.08	12.15	12.19	PM
742◈	PM												◈	12.45	12.46	12.49	12.51	12.53	12.55	12.57	12.59	1.00	1.02	1.04	1.06	1.08	1.15	1.19	PM
746◈	PM												◈	1.45	1.46	1.49	1.51	1.53	1.55	1.57	1.59	2.00	2.02	2.04	2.06		2.15	2.19	PM
750	PM													2.45	2.46	2.49	2.51	2.53	2.55	2.57	2.59	3.00	3.02	3.04	3.06	3.08	3.15	3.19	PM
754◈	PM												◈	3.45	3.46	3.49	3.51	3.53	3.55	3.57	3.59	4.00	4.02	4.04	4.06		4.15	4.19	PM
758◈	PM												◈	4.45	4.46	4.49	4.51	4.53	4.55	4.57	4.59	5.00	5.02	5.04	5.06	5.08	5.16	5.20	PM
788	PM													5.45	5.46	5.49	5.51	5.53	5.55	5.57	5.59	6.00	6.02	6.04	6.06	6.08	6.15	6.19	PM
766◈	PM												◈	7.00	7.01	7.04	7.06	7.08	7.10	7.12	7.14	7.15	7.17	7.19	7.21	7.23	7.30	7.34	PM
770	PM													9.00	9.01	9.04	9.06	9.08	9.10	9.12	9.14	9.16	9.17	9.19	9.21		9.30	9.34	PM
784◈	PM												◈	10.40	10.41	10.44	10.46	10.48	10.50	10.52	10.54	10.55	10.57	10.59	11.01		11.06	11.10	PM
700	AM													12.10	12.11	12.14	12.16	12.18	12.20	12.22	12.24		12.27	12.29	12.31	12.33	12.40	12.44	AM

◈ **Will not run Christmas Day (Dec. 25).**

The April 27, 1980 Saturdays plus Washington's Birthday and Good Friday West Chester via Media to Philadelphia (Operated by Conrail under contract with SEPTA) commuter rail service showed three trips from West Chester via Media to Philadelphia, thirteen trips from Elwyn to Philadelphia, and twenty-seven trips from Media to Philadelphia. All of the trips to Philadelphia were through trips. On Sundays and holidays, there was no service beyond Media to West Chester. (*Jose Escamilla collection*)

Philadelphia to MEDIA, ELWYN AND WEST CHESTER — SATURDAYS and Washington's Birthday and Good Friday

Train Number		PENN CENTER	30th ST. STATION	FORTY-NINTH STREET	ANGORA	FERNWOOD	LANSDOWNE	GLADSTONE	CLIFTON	PRIMOS	SECANE	MORTON	SWARTHMORE	WALLINGFORD	MOYLAN-ROSE VALLEY	MEDIA	ELWYN	WILLIAMSON SCHOOL	GLEN RIDDLE	LENNI	WAWA	DARLINGTON	GLEN MILLS	LOCKSLEY	CHEYNEY	WESTTOWN	NIELDS STREET	WEST CHESTER	
0705	AM	6.55	6.58	7.02	7.04	7.06	7.08	......	7.11	7.13	7.15	7.17	7.19	......	7.22	7.26	7.30	......	......	f 7.34	f 7.36	......	f 7.40	......	f 7.45	f 7.47	f 7.52	7.55	AM
711	AM	7.55	7.58	8.02	8.04	8.06	8.08	......	8.11	8.13	8.15	8.17	8.19	8.21	8.23	8.30	......	......	......	......	......	......	......	......	......	......	......	......	AM
1715	AM	8.55	8.58	9.02	9.04	9.06	9.08	9.09	9.11	9.13	9.15	9.17	9.19	9.21	9.23	9.27	9.32	......	......	......	......	......	......	......	......	......	......	......	AM
0715	AM	9.25	9.28	......	9.34	9.36	9.38	......	9.41	9.43	9.45	9.47	9.49	9.51	9.53	9.57	10.00	f10.02	f10.04	f10.06	f10.08	f10.10	f10.12	f10.14	f10.16	f10.18	f10.22	10.27	AM
717	AM	9.55	9.58	10.02	10.04	10.06	10.08	10.09	10.11	10.13	10.15	10.17	10.19	10.21	10.23	10.29	......	......	......	......	......	......	......	......	......	......	......	......	AM
719	AM	10.25	10.28	......	......	10.36	10.38	......	10.41	10.43	10.45	10.47	10.49	10.51	10.53	10.59	......	......	......	......	......	......	......	......	......	......	......	......	AM
1721	AM	10.55	10.58	11.02	11.04	11.06	11.08	11.09	11.11	11.13	11.15	11.17	11.19	11.21	11.23	11.27	11.32	......	......	......	......	......	......	......	......	......	......	......	AM
725	AM	11.55	11.58	12.02	12.04	12.06	12.08	12.09	12.11	12.13	12.15	12.17	12.19	12.21	12.23	12.29	......	......	......	......	......	......	......	......	......	......	......	......	PM
1727	PM	12.25	12.28	......	......	12.36	12.38	12.39	12.41	12.43	12.45	12.47	12.49	12.51	12.53	12.57	1.02	......	......	......	......	......	......	......	......	......	......	......	PM
729	PM	12.55	12.58	1.02	1.04	1.06	1.08	1.09	1.11	1.13	1.15	1.17	1.19	1.21	1.23	1.29	......	......	......	......	......	......	......	......	......	......	......	......	PM
1731	PM	1.25	1.28	......	......	1.36	1.38	1.39	1.41	1.43	1.45	1.47	1.49	1.51	1.53	1.57	2.02	......	......	......	......	......	......	......	......	......	......	......	PM
733	PM	1.55	1.58	2.02	2.04	2.06	2.08	2.09	2.11	2.13	2.15	2.17	2.19	2.21	2.23	2.29	......	......	......	......	......	......	......	......	......	......	......	......	PM
735	PM	2.25	2.28	......	......	2.36	2.38	2.39	2.41	2.43	2.45	2.47	2.49	2.51	2.53	3.00	......	......	......	......	......	......	......	......	......	......	......	......	PM
1737	PM	2.55	2.58	3.02	3.04	3.06	3.08	3.09	3.11	3.13	3.15	3.17	3.19	3.21	3.23	3.27	3.32	......	......	......	......	......	......	......	......	......	......	......	PM
739	PM	3.25	3.28	3.32	3.34	3.36	3.38	3.39	3.41	3.43	3.45	3.47	3.49	3.51	3.53	4.00	......	......	......	......	......	......	......	......	......	......	......	......	PM
1741	PM	3.55	3.58	4.03	4.05	4.07	4.10	4.13	4.15	4.18	4.20	4.23	4.25	4.27	4.29	4.31	4.35	......	......	......	......	......	......	......	......	......	......	......	PM
0743	PM	4.25	4.28	4.33	4.35	4.37	4.39	4.40	4.42	4.44	4.46	4.48	4.50	4.52	4.54	4.59	5.02	......	f 5.05	f 5.07	f 5.08	f 5.10	f 5.13	......	f 5.16	f 5.18	f 5.23	5.28	PM
779	PM	4.45	4.48	4.52	4.54	4.56	4.58	4.59	5.01	5.03	5.05	5.07	5.09	5.11	5.13	5.20	......	......	......	......	......	......	......	......	......	......	......	......	PM
781	PM	5.18	5.21	5.27	5.29	5.31	5.33	5.35	5.37	5.39	5.41	5.43	5.45	5.47	5.49	5.53	......	......	......	......	......	......	......	......	......	......	......	......	PM
1755	PM	5.45	5.48	5.52	5.54	5.56	5.58	5.59	6.01	6.03	6.05	6.07	6.09	6.11	6.13	6.17	6.22	......	......	......	......	......	......	......	......	......	......	......	PM
759	PM	6.25	6.28	......	6.36	6.38	6.40	6.42	6.44	6.46	6.48	6.50	6.52	6.54	6.56	7.04	......	......	......	......	......	......	......	......	......	......	......	......	PM
761	PM	6.55	6.58	......	7.05	7.07	7.09	7.11	7.13	7.15	7.17	7.19	7.21	7.23	7.25	7.30	......	......	......	......	......	......	......	......	......	......	......	......	PM
765	PM	7.55	7.58	......	8.04	8.06	8.08	8.09	8.11	8.13	8.15	8.17	8.19	8.21	8.23	8.29	......	......	......	......	......	......	......	......	......	......	......	......	PM
1767	PM	9.00	9.03	......	9.09	9.11	9.13	9.14	9.16	9.18	9.20	9.22	9.24	9.26	9.28	9.32	9.35	......	......	......	......	......	......	......	......	......	......	......	PM
769	PM	9.55	9.58	......	10.04	10.06	10.08	10.09	10.11	10.13	10.15	10.17	10.19	10.21	10.23	10.30	......	......	......	......	......	......	......	......	......	......	......	......	PM
785	PM	11.20	11.23	......	11.29	11.31	11.33	11.34	11.36	11.38	11.40	11.42	11.44	11.46	11.48	11.54	......	......	......	......	......	......	......	......	......	......	......	......	PM
701	AM	12.55	12.58	......	1.04	1.06	1.08	1.09	1.11	1.13	1.15	1.17	1.19	1.21	1.23	1.30	......	......	......	......	......	......	......	......	......	......	......	......	AM

SUNDAYS and New Year's, Memorial, Independence, Labor, Thanksgiving and Christmas Days

Train Number		PENN CENTER	30th ST. STATION	FORTY-NINTH STREET	ANGORA	FERNWOOD	LANSDOWNE	GLADSTONE	CLIFTON	PRIMOS	SECANE	MORTON	SWARTHMORE	WALLINGFORD	MOYLAN-ROSE VALLEY	MEDIA	ELWYN	WILLIAMSON SCHOOL	GLEN RIDDLE	LENNI	WAWA	DARLINGTON	GLEN MILLS	LOCKSLEY	CHEYNEY	WESTTOWN	NIELDS STREET	WEST CHESTER	
711	AM	7.55	7.58	8.02	8.04	8.06	8.08	......	8.11	8.13	8.15	8.17	8.19	8.21	8.23	8.30	......									......	......	......	AM
717	AM	9.55	9.58	10.02	10.04	10.06	10.08	10.09	10.11	10.13	10.15	10.17	10.19	10.21	10.23	10.29	......									......	......	......	AM
725◈	AM	11.55	11.58	12.02	12.04	12.06	12.08	......	12.11	12.13	12.15	12.17	12.19	12.21	12.23	12.29	......	NO SERVICE BETWEEN								......	......	......	PM
729◈	PM	12.55	12.58	1.02	1.04	1.06	1.08	1.09	1.11	1.13	1.15	1.17	1.19	1.21	1.23	1.29	......									......	......	......	PM
733	PM	1.55	1.58	2.02	2.04	2.06	2.08	2.09	2.11	2.13	2.15	2.17	2.19	2.21	2.23	2.29	......									......	......	......	PM
737◈	PM	2.55	2.58	3.02	3.04	3.06	3.08	3.09	3.11	3.13	3.15	3.17	3.19	3.21	3.23	3.29	......	MEDIA AND ELWYN-								......	......	......	PM
741◈	PM	3.55	3.58	4.02	4.04	4.06	4.08	4.09	4.11	4.13	4.15	4.17	4.19	4.21	4.23	4.30	......									......	......	......	PM
745	PM	4.55	4.59	5.04	5.06	5.08	5.10	5.11	5.12	5.13	5.14	5.17	5.20	5.23	5.26	5.30	......									......	......	......	PM
783◈	PM	5.55	5.58	6.02	6.04	6.06	6.08	6.09	6.11	6.13	6.15	6.17	6.19	6.21	6.23	6.29	......	WEST CHESTER ON								......	......	......	PM
761◈	PM	6.55	6.58	......	7.05	7.07	7.09	7.11	7.13	7.15	7.17	7.19	7.21	7.23	7.25	7.30	......									......	......	......	PM
765	PM	7.55	7.58	......	8.04	8.06	8.08	8.09	8.11	8.13	8.15	8.17	8.19	8.21	8.23	8.29	......									......	......	......	PM
769◈	PM	9.55	9.58	......	10.04	10.06	10.08	10.09	10.11	10.13	10.15	10.17	10.19	10.21	10.23	10.30	......	SUNDAYS AND HOLIDAYS LISTED.								......	......	......	PM
785	PM	11.20	11.23	......	11.29	11.31	11.33	11.34	11.36	11.38	11.40	11.42	11.44	11.46	11.48	11.54	......									......	......	......	PM
701	AM	12.55	12.58	......	1.04	1.06	1.08	1.09	1.11	1.13	1.15	1.17	1.19	1.21	1.23	1.30	......									......	......	......	AM

◈ **Will not run Christmas Day (Dec. 25).**

1st Ed. 100M 4-27-80 Allen, Lane & Scott, Inc. Printed in U.S.A.

The April 27, 1980 Saturdays plus Washington's Birthday and Good Friday Philadelphia via Media to West Chester (Operated by Conrail under contract with SEPTA) commuter rail service showed three trips from Philadelphia via Media to West Chester, eleven trips from Philadelphia to Elwyn, and twenty-seven trips from Philadelphia to Media. All of the trips to West Chester were through trips. On Sundays and holidays, there was no service beyond Media to West Chester. (*Jose Escamilla collection*)

2

West Chester Railroad Volunteers Restore Rail Service

On June 19, 1996, the West Chester Borough Council approved the sublease from SEPTA to 4 States Railway Service doing business as West Chester Railroad Company (WCRR) to operate 6.405 miles of track from the Borough of West Chester in Chester County Pennsylvania to Glen Mills in Delaware County, Pennsylvania. With a lot of hard work, WCRR volunteers worked over a year to clear right of way and restore the track. A former Plymouth locomotive has been placed on Market Street in downtown West Chester where the downtown station platform has been restored. The WCRR rebuilt the former Pennsylvania Railroad yard in West Chester where their locomotives, passenger coaches, and miscellaneous rolling stock are stored and maintained.

On September 21, 1997, the WCRR began excursion service, and have scheduled many special trips each year. The line passes the historic Westtown station that in 2020 is the Westtown Art Gallery featuring handcrafted art pottery. Locksley whistle stop (according to Thornbury Historical Society was established in 1875) is near one of the oldest iron slitting mill (that made nails by hand) sites in the United States and the homestead of Revolutionary War hero Colonel Persifor Frazer. In the picturesque village of Glen Mills, the Thornbury Historical Society, according to an article that originally appeared in the July 10, 1992 *Philadelphia Inquirer*, "raised $15,000 from Thornbury residents and obtained $30,000 in state funds for historic preservation" to restore the beautiful Victorian Glen Mills station. A November 2011 study "Wawa to West Chester Regional Rail Extension Ridership Forecast" by the Delaware Valley Regional Planning Commission noted that average weekday ridership on the West Chester Elwyn Line declined from 7,800 in 1979 to 4,500 in 1986 when service was discontinued west of Elwyn and continued to decline to 3,600 by 1995. By 2009, it has shown an increase to 4,900. SEPTA is currently rebuilding the line from Elwyn to Wawa to restore passenger service to Wawa.

WCRR volunteers Bob Weitzel on the left and Skip Small in the center are reopening the flange way and correcting a rail joint, because one of the joint bars (or fishplates) beneath it was loose at the Nields Street crossing in West Chester, Pennsylvania, while Guy Johnson flags the crossing to alert motor vehicle drivers in the late 1990s. As documented in the WCRHA *Newstracks* Vol. 2, Issue 3, Guy Johnson was in his train uniform because the project was taking place before and during one of the railroad's Fall Foliage trains. This is a 100 percent volunteer operation with no paid employees. (*Skip Small collection*)

Looking towards Nields Street in the borough of West Chester, Pennsylvania, the new tie in the top picture is now being inserted under the rail for the main switch. With the line out of regular service since 1986, it was necessary to replace missing rails, weathered ties, and an entire culvert south of West Chester. (*Jimmy Butler photograph*)

A gas rail cutoff saw (used by a track maintenance crew when the job has to be done immediately) is being used by Skip Small to install a replacement rail at a new culvert south of West Chester, Pennsylvania, on the WCRR line near A. Duie Pyle Inc. in March 1997. Over 100 feet of track washed out requiring replacement rail, new ties, and hardware. (*Skip Small collection*)

This is an aerial view of the West Chester, Pennsylvania, yard in 1996 before the apartments were built on the west side of the tracks. This was the original site of the Pennsylvania Railroad (PRR) roundhouse and switching yard that has been dismantled. To the right of the picture is the lot for Wyeth Pharmaceuticals which manufactured Penicillin for the war effort. (*Skip Small collection*)

WCRR Volunteers Skip Small, Steve Getz, Walt Still, and Don Callander are in the process of installing the switch on the main line at the WCRR West Chester, Pennsylvania, yard in October 1996. Before the yard work could begin, over fifty Osage orange trees had to be removed because their needle-sharp thorns were a safety hazard and can actually puncture a motor vehicle tire. (*Jimmy Butler photograph*)

Left: In this January 1997 view looking towards West Chester, a major 100-foot-long washout occurred as a result of new non railroad construction on adjacent property. This resulted in the installation of four new concrete culvert pipes as part of a massive rebuilding project. (*Jimmy Butler photograph*)

Below: It takes an amazing amount of WCRR volunteer work over a month's period to prepare the foundation and install the main switch stand as evidenced by this August 1997 scene involving WCRR volunteer track crew members Skip Small, Ricky Correll, Bud Braham, Jimmy Butler, Don Callander, and Joe Parsons. The diamond in the yard was removed and replaced by the switch, because it was no longer needed since the freight station no longer needed to be served. (*Jimmy Butler collection*)

Right: Positioning a replacement tie by means of a tie tong is an essential element in track maintenance in this July 1997 scene at the WCRR West Chester, Pennsylvania, yard. After the old tie has been removed, it is necessary to rake out the existing ballast to provide space for the replacement tie. When the replacement tie is in place, it is necessary to work the ballast under and around the tie. The manual work takes a WCRR volunteer four-person crew about one hour to complete. Everybody on this railroad has learned multiple job skills. There is an average of 3,250 wooden crossties per mile of track. (*Jimmy Butler photograph*)

Below: In 1997, John Jeppi is involved in replacing missing rail bolts on the WCRR at West Chester, Pennsylvania. Looking at the picture, the joint bar that is holding the two rails together is known as the fish plate. This particular assembly has a total of four bolts (two in each rail) to keep the rails together. (*Jimmy Butler photograph*)

Rebuilding the number 2 track at the WCRR West Chester, Pennsylvania, yard is a monumental project in April 1997. The yard had to be cleared of voluminous thorn like Osage orange trees that had an average diameter of 16 inches. Rail and ties had to be removed because of the extensive Osage orange tree root system. Half of the WCRR rail is 130-pound rail, and the remainder is 155-pound rail. (*Jimmy Butler photograph*)

WCRR locomotive No. 9 is heading a train of hopper cars being loaded with ballast at Glen Mills, Pennsylvania in the autumn of 1997. The Glen Mills quarry provided all of the ballast for the WCRR, and may have furnished all of the ballast for the original line. Ballast (a nautical term for stones used to stabilize a ship) is generally crushed stone several inches above the roadbed that supports plus keeps the railroad ties and rails in place, distributes weight evenly throughout the ties, allows for proper drainage by carrying water away from the track structure, and keeps down vegetation. (*Jimmy Butler photograph*)

In the spring of 1998, new ties are being positioned by a crane for the track switch for track three at the WCRR West Chester, Pennsylvania, yard. This was necessary to provide additional storage space for WCRR's growing fleet. All of the switches installed by WCRR volunteers were recycled from other railroads. The WCRR volunteers have become experts in organizing and completing these types of projects. (*Jimmy Butler photograph*)

In 1998, WCRR volunteers were busy relaying yard trackage at West Chester, Pennsylvania, plus servicing WCRR locomotive No. 99. The rail dates of manufacture go back to 1925. During the rebuilding, date nails from 1965 were found and saved. Hence, 1965 may have been the last year the PRR installed new ties on this main line. (*Skip Small collection*)

The summer of 1998 finds WCRR volunteers Rick Correll, Joe Giacchino (owner of 4 States Railway Service that owns the railroad), and Joe Parsons working at the West Chester, Pennsylvania, yard adding a new track beside locomotive No. 99 with the paint sprayer ready to paint No. 99. On the right side of the picture, locomotive No. 1803 is still facing north in this scene. (*Jimmy Butler photograph*)

On April 27, 1997, the WCRR operated its first work train pulled by locomotive No. 3, a type S-2 diesel electric switcher locomotive built by ALCO in October 1949. A lot of work went into rebuilding the line to make this event possible. There were a total of 1,502 type S-2 locomotives built between August 1940 and June 1950. In 1946, ALCO had 26 percent of the diesel locomotive market in the United States, and the S series of diesel switchers was well received. ALCO closed its Schenectady, New York, plant in 1969. (*Jimmy Butler photograph*)

The new railroad crossing system is being installed by Jack Brewer (a professional signal crossing installer) on the other side of Glen Mills Road in Glen Mills, Pennsylvania, in 2000. Just like a regular railroad, WCRR volunteers recognize the importance of roadway grade crossing safety. (*Skip Small collection*)

Using a hi-railer (vehicle with smaller diameter flanged wheels than on railroad cars because the wheels have to be stored to clear the ground when the vehicle is operating on a road), WCRR volunteers are removing trees at the highway crossing at Locksley in Thornbury Township, Delaware County, Pennsylvania, to improve visibility for both the railroad and motor vehicles in 2000. A wind storm caused a tree to block the rails. WCRR volunteers spend countless hours removing trees, which can sometimes delay trains. (*Skip Small collection*)

On November 15, 2000, the work train is dumping ballast for the track near Westtown in Chester County, Pennsylvania, as WCRR volunteers Walt Still, Joe Parsons, Bucky Burster, and Ernie Schwass are checking the track. As noted by WCRHA *Newstracks* Vol. 2 Issue 3, "We all got our exercise shoveling ballast." (*Skip Small collection*)

In 2000, type N-7A bay window caboose serves as the first WCRR railroad station on 250 E. Market Street in West Chester, Chester County, Pennsylvania. The WCRR website is www.wcrailroad.com and the office phone number during their operating season is 610-430-2233. The caboose was later replaced by a Plymouth locomotive, which now serves as the display for the downtown departure point. (*Skip Small collection*)

Eleven years later in 2011, the N-7A caboose sparkles in its new Tuscan red color. This caboose was rescued from the scrap yard at Modena, Chester County, Pennsylvania, and brought to the WCRR on October 15, 1996. According to information reported in the WCRHA *Newstracks* Vol. 6, Issue 4, the caboose was built by St. Louis Car Company in February 1952 for the New York Central Railroad as No. 20301. It later became Penn Central No. 21521 and Conrail No. 21521. This type of bay window construction provided better protection for the rear end crew in case of a sideswipe because by design the observer stayed inside the confines of the caboose side. A full bay approach placed the observer in a dangerous area in the event of a sideswipe collision. (*Jimmy Butler photograph*)

In this 2002 scene, a hi-rail truck (fitted with special flanged wheels that can be lowered to run on track) is pulling a generator leaving the West Chester yard heading north towards West Chester, Pennsylvania. An inspection of track by this truck is conducted over the entire line before the next run of the week. Visual inspections are important because noticing and replacing defective cross ties can avoid future track distortion issues. (*Jimmy Butler photograph*)

A replacement rail car truck for a passenger coach was brought in that was refurbished by the Strasburg Railroad in 2002. This was needed because of worn wheel flanges. Train wheels have a flange that is a projecting rim (larger diameter) on one side that keeps the train on the track. A worn wheel flange could result in the wheel rising to the top of the rail and increase the chance for a train derailment. (*Jimmy Butler photograph*)

At the WCRR West Chester, Pennsylvania yard in 2003, locomotives Nos. 1803 and 99 plus the Burro Crane (that comes in handy for a variety of projects) are awaiting their next assignment. Burro is a trade name for a series of cranes first manufactured by Cullen-Friestedt. These were self-propelled (not needing a locomotive to tow them around) and could pull a single flat car, dump car, or gondola car to carry ties, ballast, or other material. This Burro Crane was later relocated to New Castle, Delaware. (*Jimmy Butler photograph*)

Right: WCRR volunteer Derek Slifer is spray painting No. 9275 Pennsylvania Railroad baggage car with a fresh coat of Tuscan Red paint in 2019. It should be noted that other WCRR volunteers also worked on this car. (*Derek Slifer collection*)

Below: Former Pennsylvania Railroad baggage car No. 9275 has been repainted Tuscan red in this August 2001 view by WCRR volunteer Jimmy Butler. This car has been repainted multiple times by WCRR volunteers. (*Jimmy Butler photograph*)

In the summer of 2005, WCRR volunteer Kyle Chrupala is re-lettering a former Reading Railroad commuter coach No. 9114. On September 6, 2020, a number of WCRR volunteers painted coach No. 9114 Tuscan red. (*Jimmy Butler photograph*)

Standing on the platform at Glen Mills, Pennsylvania, station looking towards the Quarry Master's home are the WCRR volunteer crew. *From left to right*: Jack Nolan, engineer; Joe Wilton, conductor; Glenn Schwass, student engineer; Chuck Denlinger, former Glen Mills station master and caretaker; and Derek Slifer, brakeman are on a break in the summer of 2009. (*Ron Fricker photograph*)

Track improvements are constantly being made on the WCRR as evidenced by the track leveling project at Nields Street which was the former West Chester University station. WCRR volunteers do everything from replacing worn ties and rail to keep the line serviceable because of their sincere interest in preserving West Chester's rail heritage. In this May 19, 2005 scene looking towards Glen Mills, WCRR volunteer Skip Small is on the left. (*Jimmy Butler photograph*)

A 2005 aerial view of the WCRR West Chester, Pennsylvania, yard shows progress has been made in rebuilding the yard. (*Jimmy Butler collection*)

Left: In July 2006, to enhance public safety, both SEPTA employees and WCRR volunteers are working together to remove the catenary (overhead wire) on the WCRR line, which was formerly used for both freight and passenger service by PRR and later SEPTA. With Jimmy Butler operating WCRC locomotive No. 1803, somewhere between 15 and 20 tons of catenary was removed. On the top are WCRR volunteers Skip Small (back is shown) and Joe Parsons (in the background) on this necessary project. (*Jimmy Butler photograph*)

Below: On May 1, 2009, WCRR volunteers Frank Moscatelli and Ron Fricker are painting a former Reading Railroad commuter coach at the WCRR West Chester, Pennsylvania, yard. (*Skip Small collection*)

On May 1, 2010, WCRR volunteer Jimmy Butler is painting a WCRR diesel locomotive No. 4230 at the West Chester, Pennsylvania, railroad yard. Butler is the WCRR painting specialist and one of the many WCRR engineers. Before painting, he creates the original stencils, designs the color scheme, and applies the paint. In this view, he has applied the stencil, cut out the overlay tape, and then does the painting. (*Skip Small collection*)

Union Street in West Chester, Pennsylvania, on October 9, 2011 is the location of preparation work to install a fence to protect the WCRR railroad yard. Looking towards West Chester, WCRR volunteers are Jose Escamilla (on the flat car) and Skip Small (on the crane). Knox Equipment Rental loaned the crane for the project, and the area was being cleared to enhance safety. (*Jimmy Butler photograph*)

A WCRR hi-railer is in the West Chester, Pennsylvania, yard as WCRR volunteers Ashton Krug and Skip Small prepare to replace a tie on one of the switches in the yard on March 16, 2019. In instances of replacing a single tie, the WCRR uses the same type of equipment and tools that were used by the Pennsylvania Railroad before automation. (*Steve Krug photograph*)

In the aerial view of the WCRR West Chester, Pennsylvania, yard, note the load of new ties shown in the lower right-hand portion of the picture that were delivered and being installed in the spring of 2020 by WCRR volunteers. (*Andrew L. Keates drone photograph*)

3

West Chester Railroad Locomotives

The West Chester Railroad Company (WCRR) has acquired a number of locomotives summarized in the below paragraphs. Electro-Motive Division of General Motors Corporation will be shown as EMD. American Locomotive Company will be shown as ALCO. Montreal Locomotive Works will be shown as MLW. For both ALCO and MLW, Century such as in Century 424 will be shown as C424. General Electric Company will be shown as GE.

WCRR No. 1 is a 20-ton Plymouth switcher built in April 1941 for the U.S. Army Corps of Engineers. It was originally gas powered; however, has been re-motored with a diesel. It was donated by the Glen Mills quarry where it worked. It currently is on display on Market Street in downtown West Chester where the downtown station platform has been restored.

WCRR No. 3 is a 1,000-horsepower type S2 diesel electric switcher locomotive built by ALCO in October 1949 as No. 5026 for the Chesapeake & Ohio Railway and later became No. 9115 for the Baltimore & Ohio Railroad. It was acquired by Rohm & Haas and became No. 3. After it was damaged by freezing, it was purchased by the Wilmington & Western Railroad. It was acquired by the WCRR in exchange for an SW600 provided to the Wilmington & Western Railroad and is currently not in service.

WCRR No. 9 is a 400/550-horsepower type 65 tonner diesel electric locomotive built by GE in April 4, 1941 as USAX No. 7079.

WCRR No. 1315 is a four-axle 2,000-horsepower type C420 diesel electric switcher built by ALCO for the Louisville & Nashville Railroad in September 1966. This is reported to be the last ALCO locomotive of any type built for the railroad. Following retirement on July 20, 1982, it was sold to Chrome Crankshaft, later went to Indiana Hi-Rail, became Wabash & Ohio No. 315, and RMW Ventures No. 315. It was extensively restored back to running condition by Joseph Giacchino who has a dedicated passion for history. It was used on the WCRR for two years. Since it was important for historic reasons for this locomotive to be returned to its original operational area, it was traded in for No. 4230. At the Oak Ridge, Tennessee, location of the Southern Appalachia Railway Museum, No. 1315 was rebuilt, repainted, and debuted on September 21, 2008 at the Louisville & Nashville History Society meeting.

WCRR No. 1803 is a 1,800-horsepower diesel electric locomotive built in April 1958 by Montreal Locomotive Works (MLW) as a type RS-18 No. 8762 for the Canadian Pacific Railway (CPR). There were seventy-two of these locomotives (Nos. 8729–8800) built for CPR. It was rebuilt by CPR as a type RS-18u and renumbered 1803. WCRR acquired this locomotive in 1998. This locomotive has only had two owners CPR and WCRR.

WCRR No. 4213 is a four-axle 2,400-horsepower type C424 diesel electric locomotive built in March 1965 by MLW for the Canadian Pacific Railway. It was sold to Quebec Railway Corporation and renumbered 4202. It was acquired by the WCRR and arrived on October 1, 2010. WCRR restored the original No. 4213.

WCRR No. 4230 (the railroad's newest locomotive) is a four-axle 2,400-horsepower type C424 diesel electric locomotive built in May 1965 by MLW for the Canadian Pacific Railway. It was sold to the New Brunswick East Coast Railway in December 1998 and was acquired by the WCRR in 2005.

WCRR No. 6499 is a four-axle 1,750-horsepower type GP9 road switcher locomotive built by EMD in October 1956 for the Baltimore & Ohio Railroad. It was acquired by the Youngstown & Austintown Railroad and later by the Ohio Central where it was renumbered 99. In 1997, the WCRR acquired the locomotive. With help from the Pennsylvania Railroad Technical & Historical Society Philadelphia Chapter, it was repainted into a Pennsylvania Railroad paint scheme. The locomotive has been repainted and renumbered 6499.

WCRR No. 7706 is a four-axle 2,000-horsepower type GP38 diesel electric locomotive built in 1969 by EMD for the Penn Central Transportation Company (Penn Central). Following the bankruptcy of Penn Central, it went to Consolidated Rail Corporation (Conrail) in 1976. After it was purchased by PECO Energy in the mid-1990s, it was later transferred to the Exelon Corporation and worked at the Cromby Generating Station in Phoenixville. The locomotive was donated to the WCRR in 2011.

All the painting for the railroad's locomotives was done or led by Jimmy Butler. All major mechanical and electrical work on the locomotive engines was done by Joseph Giacchino. Based on WCRR experience, Alco locomotives are not only valuable from an historical standpoint, but also have great pulling power, more reliable, and are easier to repair. All heavy equipment moves were done by Mc Hugh Locomotive & Equipment.

On May 4, 1997, WCRR locomotive No. 9 is pulling the former New York Central Railroad caboose No. 642 for the WCRR's first work train run. In this view, the caboose had been repainted a maroon color. WCRR volunteer Joe Parsons was the main electrician until he passed. (*Jimmy Butler photograph*)

WCRR locomotive No. 99 and Ohio Central Railroad System No. 3217 are in route in Ohio to West Chester, Pennsylvania, around the summer of 1997. Type GP40 diesel electric road switcher No. 3217 rated 3,000-horsepower was built by EMD in November 1968 for Penn Central Transportation Company. It was acquired by the Buckingham Branch Railroad and later by the Ohio Central Railroad System. (*Joseph Giacchino collection*)

WCRR No. 99 is backing up to pick up a passenger coach for a trip at the West Chester, Pennsylvania, yard in the spring of 1999. WCRR volunteers Walt Still was the engineer on his favorite locomotive, and conductor Glen Schwass was on the locomotive platform using hand signal to guide the locomotive into position. (*Jimmy Butler photograph*)

With a blanket of snow on the ground, WCRR locomotives Nos. 99 and 3 are basking in the sunshine of a winter day in 2000 at the West Chester, Pennsylvania, yard. (*Jimmy Butler photograph*)

In July 2000, WCRR diesel locomotive No. 9 is at the West Chester, Pennsylvania, yard to switch around coaches to facilitate the installation of new doors for the coaches and dining car. The new doors were designed and hand crafted by WCRR volunteer Ricky Coreyell. Within a short time, a thunderstorm hit the area. (*Jimmy Butler photograph*)

On a sunny day in 2002, WCRR type S-2 Alco built locomotive No. 3 is on track 3 at the West Chester, Pennsylvania, yard. From an historical standpoint, the type S-2 locomotive was used on yard services and freight operation on branch lines. This was the third locomotive purchased by the WCRR. (*Jimmy Butler photograph*)

Above: WCRR locomotives Nos. 99 and 9 are at the West Chester, Pennsylvania, yard basking in the bright sunshine of a winter day in 2000. This was an expensive railroad to build because of the deep valleys and many streams it crossed. By 1858, rails were being placed from Wawa to West Chester, and the first train from Philadelphia reached West Chester on November 11, 1858. (*Skip Small collection*)

Right: An inside view of the controls of WCRR locomotive No. 99 brings to mind the training required to operate the locomotive along with the maintenance requirements to keep it in service. All of this is done by the competent WCRR volunteers. (*Skip Small collection*)

In October 2004, WCRR volunteer Skip Small is standing in the grade taking a painting break on locomotive No. 6499. The WCRR volunteers of this railroad spend an enormous amount of time throughout the year maintaining track, checking the right of way which includes cutting trees that have fallen on the track after a storm, and making sure roadway signal crossing devices are properly functioning. Hours of preparation work are completed plus the locomotive and rolling stock are checked before the train leaves the railroad yard for the station. (*Jimmy Butler photograph*)

On October 4, 2000, WCRR locomotive No. 6499 is pulling vintage former New York Central Railroad bay window caboose. (*Jimmy Butler photograph*)

Above: A WCRR work train headed by locomotive No. 6499 and staffed by WCRR volunteers is making its way thru tranquil wooded Oakbourne Park returning from a tree cutting session that was necessitated by a storm. The caboose was used on the train to provide needed height to remove high tree limbs so that there was clearance between tree branches and the train cars. (*Andrew L. Keates photograph*)

Left: On October 23, 2019, a WCRR excursion train headed by locomotive No. 6499 is at the historic Glen Mills, Pennsylvania, station with the picture taken from the Quarry Master's side. On the right side of the locomotive is the white post for hitching horses. (*Steve Rogers photograph*)

Locomotive No. 1315 is on a test run in 2004. This type Alco C420 locomotive was the last C420 built for the Louisville & Nashville Railroad (L&N). The L&N operated under the same name continuously for 132 years surviving the Civil War plus economic depression and became a 6,000-mile system. Throughout its lifetime, it was economically strong and earned the name "The Old Reliable." In 1982, it was absorbed by the Seaboard Coast Line, which became part of CSX Transportation. (*Jimmy Butler photograph*)

Canadian Pacific Railroad diesel locomotive No. 1803 is handling a freight train in Canada. This locomotive was later purchased by the WCRR. The Canadian Pacific Railway, incorporated on February 16, 1881, completed its transcontinental route across Canada on November 7, 1885. It expanded its rail network in the United States in 1990 by acquiring the Minneapolis, St. Paul & Sault Ste. Marie Railroad plus the Delaware & Hudson Railway to give it access to ports in the northeast part of the United States. (*Jimmy Butler collection*)

In August 1998, former Canadian Pacific Railroad diesel locomotive No. 1803 is at the West Chester, Pennsylvania, yard. This locomotive was later repainted to West Chester colors by WCRR volunteer Jimmy Butler. (*Jimmy Butler photograph*)

WCRR locomotive No. 1803 is heading a lineup of WCRR locomotives and rolling stock at the West Chester, Pennsylvania, yard. (*Andrew L. Keates photograph*)

In 1999, a test is made for double header operation of WCRR diesel locomotives Nos. 99 and 1803. Locomotive No. 1803 was repainted from the Canadian Pacific Railroad paint scheme to the WCRR paint scheme. (*Jimmy Butler photograph*)

In late August 2001, WCRR diesel No. 1803 is sitting on track two at the West Chester, Pennsylvania, yard before hooking up to passenger coaches to head to West Chester, Pennsylvania, for a picnic run. (*Jimmy Butler photograph*)

In March 2000, this was the last day for locomotive No. 1803 to have its short hood end facing north. The next day, the locomotive was taken to Philadelphia Zoo wye to be reversed so that it was faced in the correct direction coming into West Chester. (*Jimmy Butler photograph*)

In 2007, WCRR locomotive No. 1803 is at the West Chester, Pennsylvania, yard ready for the next assignment. On the right, WCRR locomotive No. 4230 is in the background. (*Jimmy Butler photograph*)

The WCRR work train, powered by locomotive No. 1803, is passing the A. Duie Pyle Inc. on its way back to the West Chester, Pennsylvania, yard after WCRR volunteers did track maintenance work in July 2006. (*Jimmy Butler photograph*)

Passing through the Wiggins Christmas Tree Farm area, WCRR locomotive No. 1803 is leading a passenger excursion heading back to West Chester. At Christmas time, the WCRR train stops at the Wiggins Christmas Tree Farm to allow train passengers to pick out and cut their own tree. WCRR volunteers load each passenger's tree on to a railroad flat car attached to the train. At the West Chester station, WCRR volunteers unload each customer's tree. Customers are then responsible to secure their tree for the trip home. (*Jimmy Butler photograph*)

On June 19, 2011, WCRR locomotive No. 1803 looks nice in the new experimental paint design with "Radio Equipped" noted below the cab window at the West Chester, Pennsylvania, yard. All of the painting is done outside, which means that weather conditions for any painting day are carefully reviewed. (*Jimmy Butler photograph*)

Three WCRR locomotives headed by No. 4213 are going under the U.S. Highway Route 202 Bypass on a return trip to West Chester, Pennsylvania. A triple header is a rare event on the WCRR. (*Jimmy Butler photograph*)

Canadian Pacific Railways locomotive No. 4230 is leading a freight train in Canada. The locomotive was in Canada until 1998, and the picture was taken before being acquired by the WCRR. This locomotive was added to the WCRR fleet because of its hauling capacity, fuel efficiency, and weather conditioning already in place. (*Jimmy Butler collection*)

Former Canadian Pacific Railways locomotive No. 4230, looking weathered, features the reporting mark WCRL in this view at the Lenni Yard in 2005. A reporting mark is an alphabetic code of two to four letters used to identify owners of railroad rolling stock. Hence, the car can be tracked by the railroad it is on and shared with other railroads and customers. (*Jimmy Butler photograph*)

In 2007, Jimmy Butler is the engineer on WCRR locomotive No. 4230 with a Pennsylvania Railroad decal on the front is at the West Chester, Pennsylvania, yard. The locomotive was painted black and decaled to resemble a Pennsylvania Railroad paint scheme, because the Pennsylvania Railroad controlled all railroad service to West Chester by the end of 1881 until the Pennsylvania Railroad and New York Central Railroad merged into the Penn Central Transportation Company in 1968. (*Jimmy Butler collection*)

At the West Chester, Pennsylvania, yard, locomotive No. 4230 is being repainted. Many WCRR volunteer hours are spent on the necessary preparation work before any painting can be done. All painting is done outside at the West Chester, Pennsylvania, yard. Wind, humidity, temperature, chance of precipitation, and availability of WCRR volunteers determine when painting can be done. (*Jimmy Butler photograph*)

On May 15, 2010, WCRR volunteer effort has resulted in a beautiful paint scheme for WCRR locomotive No. 4230 at the West Chester, Pennsylvania, yard. The metal press and vice at the work area between the two locomotives are extra heavy duty for some of the metal work done by WCRR volunteers. The wooden boardwalk in front of both locomotives defines where each locomotive should be in the yard and for WCRR volunteers to cross safely between tracks. (*Jimmy Butler photograph*)

Locomotives Alco type C424 No. 4230 and Montreal Locomotive Works type RS-18 No. 1803 look neat in the well-designed new paint scheme at the West Chester, Pennsylvania, yard. There were 190 type C424 locomotives produced from April 1963 to May 1967, and 351 type RS-18 locomotives were produced from December 1956 to August 1968. The locomotive number boards are all hand designed to fit the designated slot. (*Skip Small collection*)

Matching paint schemes on WCRR Alco type C424 locomotives Nos. 4230 and 4213 at A. Duie Pyle Inc. in West Chester, Pennsylvania, represent a tremendous amount of work by WCRR volunteers who have given this railroad a new future. This is a rare double header on the WCRR. Alco locomotives have an excellent reputation for fuel economy; however, they have high maintenance demands, which is good for short trips where they are back to the West Chester yard for inspection. (*Jimmy Butler photograph*)

In 2005, a car manufactured by the Budd Company is at the West Chester, Pennsylvania, yard before movement to the current location where it is the railroad's office. This required additional yard trackage that was completed by Skip Small, Larry Jensen, Andrew L. Keates, Jacob A. Keates, and other WCRR volunteers in the fall of 2018. The flat car at the left-hand side of the picture is used by maintenance of way WCRR volunteers to transport railroad ties and other equipment for track work. (*Jimmy Butler photograph*)

WCRR volunteer engineer Derek Slifer is in the process of checking the engine coolant water level and the governor oil level for locomotive No. 4230 in October 2012. These inspections are similar to those done on a major railroad. (*Derek Slifer collection*)

There are a lot of special skills needed in maintaining a locomotive as evidenced by WCRR volunteer Derek Slifer rebuilding a locomotive water pump for locomotive No. 7706 in June 2015. It took Derek Slifer and other WCRR volunteers over a day to remove, rebuild, and reinstall the water pump that weighed over 100 pounds. (*Derek Slifer collection*)

Right: The replacement of a piston on locomotive No. 4230 in November 2012 required the team effort of WCRR volunteers. *From left to right*: Derek Slifer and Joseph Giacchino with Jimmy Butler on top of the locomotive. (*Derek Slifer collection*)

Below: On March 3, 1990 at Jackson, Michigan, locomotive No. 7706 is in its Conrail paint scheme. This 2,000-horsepower type GP38 diesel electric locomotive was built by EMD at LaGrange, Illinois, on July 10, 1969 for Penn Central Transportation Company (Penn Central). Locomotive weight on trucks was 263,473 pounds. After the bankruptcy of Penn Central, it went to Consolidated Rail Corporation (Conrail) in 1976. (*R.E. Holmes photograph, Derek Slifer collection*)

A Norfolk Southern Railway (NS) local freight train picked up No. 7706 at the Cromby Generating Station in Phoenixville, Pennsylvania, on October 17, 2012 and is shown between NS locomotives No. 3001 (3,000-horsepower type GP40-2 built by EMD in December 1973) and No. 9071 (4,400-horsepower type D9-44CW built by General Electric Company in March 1997) passing the former Reading Railroad Valley Forge, Pennsylvania, train station on its way to the WCRR in West Chester, Pennsylvania. (*Derek Slifer photograph*)

WCRR type GP-38 locomotive No. 7706 is undergoing restoration and repainting by WCRR volunteers in this August 8, 2015 view at the West Chester, Pennsylvania, yard. This was No. 7706's first complete repainting. Before the repainting, all mechanical plus electrical maintenance along with rust removal was performed by WCRR volunteers. EMD built 706 GP38 locomotives between January 1966 and December 1971. (*Derek Slifer photograph*)

From left to right: WCRR volunteers Derek Slifer and Dave Kent are replacing a water pump on WCRR locomotive No. 7706 in June 2015. This locomotive has two water pumps each mounted directly to the engine block with ten bolts. (*Derek Slifer collection*)

Refurbished and repainted WCRR locomotive No. 7706 looks perfect in this picturesque scene at the Glenn Mills, Pennsylvania, station. The luggage cart out in the front was once used for mail and express by Pennsylvania Railroad porters. Across the tracks, the Quarry Master's home (128 Glen Mills Road) has been carefully restored by the Byrsa Bistro, which is a dining establishment in 2020. (*Terry Redeker photograph, Derek Slifer collection*)

In 2015, McHugh Locomotive & Equipment uses a 40-ton crane to remove the failed main generator from locomotive No. 6499. At the McHugh shop, it was found that the main generator armature shaft was damaged at the bearing location, which necessitated machining it undersize, welding that area larger than the required size, and re-machining the bearing area to the required outside diameter. The armature electrical defects were repaired, and armature was reconditioned. All of the field and interpole windings from the stator case were removed, failed windings were repaired, reassembled back into the stator case, were electrical tested, and main generator was completely reassembled. The auxiliary main generator was repaired and reconditioned. (*Jimmy/J. C. McHugh collection*)

With the main generator removed, McHugh Locomotive & Equipment repaired cracked and rotted areas of the locomotive cab to prevent water from leaking into the electrical cabinet. The reconditioned main generator and component parts were delivered, the 40-ton crane was used to install the reconditioned main generator, and the project was completed so the WCRR could use locomotive No. 6499 for the 2015 Santa Claus trains. (*Jimmy/J. C. McHugh collection*)

McHugh Locomotive & Equipment is using a 150-ton hydraulic truck to lift locomotive No. 1803 on July 14, 2016. On the left is Matt Urban and on the right is Dave Brady. The truck assembly was removed from under No. 1803 by the McHugh service crew and taken to the McHugh shop where it was completely disassembled. Following removal of the good wheels and bull gear from the failed axle, a new axle was manufactured and modified to accept Timken 7½-inch by 12-inch roller bearings. The two-axle truck frame journal boxes were modified to accept the Timken roller bearings. Following checking of the used wheels and bull gear, they were placed on the new axle set. Wheels on the new and other axle set were reprofiled so that all diameters matched. Following minor repairs made on the truck assembly and traction motors, the truck assembly was reassembled. (*Jimmy/J. C. McHugh collection*)

The McHugh Locomotive & Equipment crew returned to the WCRR with the truck assembly, plus another 150-ton crane arrived which picked up locomotive No. 1803 allowing the McHugh crew to install the truck assembly back under the locomotive. On the left is Steve Urban, and on the right is Matt Urban with his back facing us, J.C. McHugh with his arm facing out, and Jimmy McHugh behind him. The second day, the McHugh crew completed their work. Final minor maintenance work was completed by WCRR volunteers, and No. 1803 went back into service. (*Jimmy/J. C. McHugh collection*)

WCRR locomotive No. 4230 is about to cross a bridge at Oakbourne Park in this scenic location of Westtown Township in Chester County, Pennsylvania. This railroad provides a scenic trip for a rapidly growing part of southeastern Pennsylvania. (*Ron Fricker photograph*)

On June 15, 2019, amid a dense growth of trees at this rural railroad crossing near the Locksley station shelter, WCRR locomotive No. 7706 is handling a freight train photo charter for rail fans that was arranged by Andrew L. Keates. In March 2020, a tree fell on the shelter damaging the entire roof as well as cracking one of the supporting beams. By September 2020, all repairs to the shelter were completed with financial assistance from Thornbury Township. (*Andrew L. Keates photograph*)

4

West Chester Railroad Passenger Service

From the station located at 250 East Market Street in historic downtown West Chester, Chester County, Pennsylvania, the West Chester Railroad Company (WCRR) operates 6.405 miles of track via Westtown station (in 2020 Ken Kazanjian's Westtown Station Gallery), Cheyney station (home of Cheyney University), Locksley station shelter (near an iron milling site), and Glen Mills, Delaware County, Pennsylvania, with its restored Victorian station. In the Greater Philadelphia area, Cheyney has the only operating United States Post Office on a SEPTA historical rail line. This was once the Pennsylvania Railroad central division on the West Chester and Phoenixville branch. The WCRR is owned by the 4 States Railway Service, Inc. and is operated by the West Chester Railroad Heritage Association, a non-profit organization dedicated to the preservation of the railroad. Excursion passenger service between West Chester and Glen Mills began on September 21, 1997. All employees of the railroad from the train crew to the track crew are dedicated WCRR volunteers. There is always room for more dedicated volunteers, and no prior railroad experience needed. Even when trains are not running, the crew is constantly keeping the right of way clear of fallen trees, maintaining track, locomotives, passenger coaches, and equipment. The railroad has acquired and refurbished locomotives and passenger coaches so that it can operate passenger service in an accurate manner for the education and enjoyment of the public. Special excursions are operated nine months of the year on specific weekends for a ninety-minute journey back in time when the railroad was the main way to travel. For example, in 2019, the railroad operated an Easter Bunny Express, Mother's Day Express, Memorial Day Special, Father's Day Express, History Day, Summer Picnic Special, Fall Foliage, Great Pumpkin Express, Trains & Soldiers plus author book signing event, Santa's Express, and Christmas Tree Train.

On a snowy December 1960, a Pennsylvania Railroad two-car train, of MP54 (M is for motor, P is for passenger) commuter coaches, is at the Clifton-Aldan, Pennsylvania, station on the Media-West Chester line. In 1951, there were 481 MP54 cars of all types in service. They were used from 1915 to 1981 on Pennsylvania Railroad, Penn Central, Conrail, and later SEPTA and New Jersey Transit. (*James Gillin photograph*)

Reading Railroad commuter coach No. 9124 is part of a two-car train at Reading Terminal in Philadelphia. These two cars were part of the Reading's original fleet of 100 olive green cars built between 1931 and 1932 at the Harlan & Hollingsworth plant in Wilmington, Delaware, that had been acquired by Bethlehem Steel in 1904. In 1964, thirty-eight cars of the original fleet were rebuilt by the Reading with a blue and white paint scheme, new seat cushions, fluorescent lights, and better ventilation and became known as "Blueliners." They were used by SEPTA until 1990. Five of these coaches (Nos. 9107, 9109, 9114, 9117, and 9124) were acquired by WCRR. (*Jimmy Butler photograph*)

Glen Mills, Pennsylvania (southern terminus of WCRR passenger operation), is the location of type S-2 diesel electric switcher locomotive No. 3 and type GP9 diesel electric locomotive No. 99 on May 23, 1997. In this view, Glen Mills station is on the left and the Quarry Master's house is on the right. Note, the catenary is still in place, and the locomotives are facing south. (*Jimmy Butler photograph*)

On November 30, 1997, WCRR type S-2 diesel electric switcher locomotive No. 3 is at the West Chester yard waiting for the proper time to go north to the West Chester, Pennsylvania, station. This locomotive was originally No. 9115 built by ALCO in 1949 for the Baltimore & Ohio Railroad. It was renumbered #3 by Rohm & Haas, purchased by the Wilmington & Western Railroad where it was rebuilt, and in 2012 was traded to the WCRR in exchange for another locomotive. (*Skip Small collection*)

Two former Reading Railroad multiple unit passenger coaches (last used by the Southeastern Pennsylvania Transportation Authority [SEPTA]) are at Jim Thorpe, Pennsylvania. These outstanding cars are now used by the WCRR on their West Chester–Glen Mills line. Under SEPTA, the former Pennsylvania Railroad and Reading Railroad commuter lines have been carefully modernized. (*Skip Small collection*)

Above: With the electric overhead catenary still in place around the year 2000, WCRR type S-2 locomotive No. 3 is preparing for a departure. The type S-2 locomotive could produce 69,000 pounds of starting tractive effort and about 30,000 pounds of continuous tractive effort plus was able to negotiate sharp curves and tight clearances. (*Joseph Giacchino collection*)

Left: In the West Chester, Pennsylvania, yard, WCRR locomotive No. 99 is ready for the first trip on September 21, 1997. This was a special run with the flowers and bow on the front of the locomotive from a wedding ceremony held at Glen Mills station. This was the first wedding held on the railroad. Ann Giacchino handled all of the catering arrangements for the wedding reception at Glenn Mills. (*Skip Small collection*)

On September 21, 1997, Joseph Giacchino, chief executive officer and owner of Four States Railway Service, Inc., is helping the crew on the railroad's first passenger train at Glen Mills, Pennsylvania. Joseph Giacchino is passionate about his locomotives doing maintenance and mechanical work himself with the help of WCRR volunteers. In addition, he is the engineer on many of the railroad's trains. (*Skip Small collection*)

West Chester, Pennsylvania, station in 1997 is the location of No. 9, a 65-ton diesel locomotive built by General Electric Company on April 4, 1941 as USAX No. 7079 (engine facing in the direction of Glen Mills). This locomotive was used at an arsenal until about 1963. It was purchased as U.S. Army surplus by the Black River & Western Railroad at Ringoes, New Jersey, where it powered the work train and light freight service. In 1978, it was purchased by the Octoraro Railway Inc. at Kennett Square, Pennsylvania, and became No. 9. The locomotive had several leases and in 1999 was donated to the Hockessin Area Heritage Association (predecessor of the WCRHA). This locomotive was saved from scrapping with a tremendous amount of effort. In 2020, No. 9 is officially retired; however, it is still part of the railroad's fleet. (*Skip Small collection*)

In the late summer of 1998, former Canadian Pacific locomotive No. 1803 is at the West Chester, Pennsylvania, yard with three former Reading Railroad multiple unit commuter coaches and former Pennsylvania Railroad baggage car No. 9275. This heritage railroad by definition has recreated a former railroad line providing a genuine education experience as to what rail transportation was like before widespread use of the automobile. (*Jimmy Butler photograph*)

In 1998 at the West Chester, Pennsylvania, yard, WCRR diesel locomotive No. 99 with former PRR baggage car No. 9275 built in the 1920s and a former Reading Railroad multiple unit commuter coach are all on main line on the right side of the picture and a former Reading Railroad commuter coach is on track two towards the left side of the picture. (*Skip Small collection*)

Locksley station shelter (located just north of the Locksley Road crossing in Delaware County, Pennsylvania) finds a WCRR train at this scenic location in 1998. Note the only remaining Pennsylvania Railroad signal mask on this railroad is in the center of the picture. In the spring of 2011, the Locksley whistle stop rest area was created and designed by Scott Quinn for his Eagle Scout project, which was approved and completed with the assistance of his Boy Scout Troop 222 of Westtown, Pennsylvania. (*Skip Small collection*)

In September 1998, WCRR locomotive No. 99 is heading a nicely painted train. The WCRR is a rolling railroad heritage excursion that has reintroduced the public to what was once the prime mode of transportation before the advent of the automobile age. (*Jimmy Butler photograph*)

WCRR volunteers Walt Still and Jimmy Butler are on the front platform of locomotive 99 for its first run in its new coat of paint on September 18, 1998. A lot of work takes place on the day's first passenger run. The engineer starts the locomotive engines. Each day's operation is set up by the conductor. The conductor helps the engineer put the train together. Air brakes are checked by the engineer. Conductor and engineer work together for the brake test. All train orders are given by the conductor. There is a crew meeting before departure to West Chester to cover needed assignments. (*Deven Butler photograph*)

WCRR locomotive No. 1803 is moving with three coaches and a baggage car at the West Chester, Pennsylvania, yard in 1999. Each WCRR passenger train has at least one engineer, one conductor, one brakeman who assists the conductor, one trainman who collects tickets and assists passengers getting on and off the train, and one extra person who handles whatever else is needed. (*Jimmy Butler photograph*)

With vintage automobiles in the parking area as part of the "Great Train Robbery" event, WCRR locomotive No. 99 is at the picturesque Glen Mills, Pennsylvania, station on May 22, 1999. This is the Chester Creek Historic District that was listed in the National Register of Historic Places on March 24, 1972. The Glen Mills station contributes to the historical and architectural significance of the district. (*Skip Small collection*)

SEPTA No. 50 Brookville Locomotive type BL1500 plus WCRR locomotive No. 99 are seen in this fall view at the Glen Mills, Pennsylvania, station in 1999. (*Skip Small collection*)

With a layer of snow on the ground at Glen Mills, Pennsylvania, station in 1999, WCRR locomotive No. 99 is waiting for departure time. In 1881, this line became the West Chester branch of the Pennsylvania Railroad. The station, shown on the right-hand side of the picture was built during 1880–1881. On the left side of the picture, the building was the original Quarry Masters home and in 2020 is the Byrsa Bistro restaurant. (*Skip Small collection*)

In 1999, WCRR locomotive No. 99 is at the West Chester, Pennsylvania, station. Note the creative paint scheme used to highlight West Chester on the locomotive. This heritage railroad is a museum of moving history with its vintage rolling stock and historic diesel locomotives. (*Skip Small collection*)

Around 1999, WCRR locomotive No. 1803 is crossing under a privately owned bridge that is no longer there, having been removed for safety reasons at this location near A. Duie Pyle Inc. in West Chester, Pennsylvania. The type RS-18 locomotive was built by Montreal Locomotive Works in April 1958 for the Canadian Pacific Railway and was acquired in 1998. (*Skip Small collection*)

On October 21, 2000, a scheduled WCRR evening "Murder Mystery Train" is at the Glen Mills, Pennsylvania, station with WCRR volunteer trainman Ernie Taggart ready to assist passengers for boarding the train. (*Skip Small collection*)

Remnants of snow in the bright sunshine of a 2000 winter day add to the delightful scene of a WCRR train powered by locomotive No. 99 at the Glen Mills station. The unincorporated community of Glen Mills is in Concord Township in Delaware County, Pennsylvania. Glen Mills was named for two paper mills (one built in 1835 and the second one built in 1846). Between 1864 and 1878, these mills supplied the United States government with special paper for the printing of government bonds. Although these mills are no longer standing, an earlier grist mill (a building with some machinery in it) survives built by Nathaniel Newlin in 1704 and known as the Newlin Grist Mill is part of a park in Glen Mills. (*Skip Small collection*)

Passengers are waiting to board the WCRR train powered by locomotive No. 99 at Glen Mills, Pennsylvania station in 2000. The WCRR contributes to the economic vitality of West Chester-Glen Mills corridor of Chester County. Visitors riding the train get a chance to see the scenic beauty of the area and board the train in vibrant downtown West Chester. The WCRR serves as an educational attraction for people of all walks of life making a positive contribution to Chester and Delaware County. (*Skip Small collection*)

Former Reading Railroad multiple unit passenger coach No. 9107 is part of the train at Glen Mills, Pennsylvania, in this 2000 winter scene. The significance of the platform bricks was that it was a station renovation fundraiser by the Thornbury Historical Society. Each brick is engraved with a different name. In the spring of 2020, the area around the station was planted by WCRR volunteers including the Keates family with over seventy-six bulb gardens in honor of COVID-19 victims. In addition, garden arrangements were planted at all of the other WCRR stations. The bulbs were donated by Wedgewood Gardens of Glen Mills, Pennsylvania. (*Skip Small collection*)

On March 18, 2000, WCRR passenger train headed by locomotive No. 1803 is at the West Chester, Pennsylvania, station ready for a trip to Glen Mills and back. There is ample free parking plus a children's playground here. In 2020, West Chester is known for its restaurants including nearby Lorenzo's Steaks & Hoagies at 216 Market Street. (*Skip Small collection*)

A crowd of passengers of all ages are boarding the WCRR train in downtown West Chester, Pennsylvania, powered by locomotive No. 99 in 2000. Tickets can be purchased online at www.wcrailroad.com or at the station located at 230 E. Market Street. Each of the railroad's passenger coaches holds about seventy-five guests. Train riders get a chance to experience the railroad's history, which provides for a positive educational learning experience. (*Skip Small collection*)

WCRR locomotive No. 99, with a former Pennsylvania Railroad baggage car and former Reading Railroad passenger coach, have passengers ready to board the train at West Chester, Pennsylvania, in 2000. Note the overhead catenary was still in place. At one time, this station platform had a roof to protect passengers from inclement weather. To the right of the station platform is Railroad Street, which is the only surviving cobblestone street in West Chester, Pennsylvania. (*Skip Small collection*)

Countless hours of work by dedicated WCRR volunteers made possible this WCRR train powered by locomotive No. 3 loading passengers at West Chester, Pennsylvania, for the excursion to Glen Mills and return in 2000. For example, an August 4, 2020 rain and wind storm resulted in eight trees falling on to the railroad right of way with three of the trees covered in vines, and they were cleared before the train ran. (*Skip Small collection*)

A southbound WCRR train headed by type GP9 locomotive No. 99 is crossing Union Street in West Chester, Pennsylvania, on a trip to Glen Mills in 2000. GP stands for general purpose. Between 1954 and 1959, EMD produced 3,449 GP9 locomotives. Known as a practical, useful, and reliable diesel locomotive, the Chesapeake & Ohio Railway had 363, Illinois Central Railroad had 348, Norfolk & Western Railway had 306, and the Pennsylvania Railroad had 270. (*Skip Small collection*)

With snow on the ground, WCRR locomotive No. 1803 along with former Reading commuter coach No. 9124 are ready for the next trip. Wintertime runs are fun for all ages. On a cold day, nothing compares with the sound of a train on the track and looking out the frosty windows viewing the beautiful countryside. (*Skip Small collection*)

On May 17, 2001, WCRR baggage car No. 7551 (later changed to No. 9275) was dedicated to Pete Small, the father of Skip Small. The baggage car was privately owned by Skip Small until March 2020 when he donated it to the West Chester Railroad Heritage Association. (*Skip Small collection*)

West Chester, Pennsylvania, yard in July 2000 is the location of former Reading Railroad "Blueliner" commuter coach No. 9124, which has had its traction motors and pantographs removed. (*Jimmy Butler photograph*)

Passengers arriving at the Cheyney station (located at 119 Station Road in Cheyney, Pennsylvania) had a nice group of vintage automobiles to view before their return trip to West Chester in 2001. Currently leased to the United States (U.S.) Post Office, this is the only operating U.S. Post Office on a heritage railroad on the SEPTA system. The 2008 thriller movie *The Happening* was filmed at this station that was renamed Filbert Station for that movie. Under U.S. 111th Congress Public Law 119, this station became the Captain Luther H. Smith U.S. Army Air Forces Post Office on June 9, 2010. Captain Luther H. Smith was a World War II pilot who flew 133 combat missions over Europe. (*Jimmy Butler photograph*)

In August 2001, brightly repainted WCRR former Reading Railroad "Blueliner" commuter coach No. 9114 is ready for the next run at the West Chester, Pennsylvania, yard. Coach No. 9114 was repainted again on August 30, 2020 by WCRR volunteers Derek Slifer, Jose Escamilla, Tyler Haney, Beth Keates, Louie Montanaro, Dan Steil, and Marc Zimmerman. (*Jimmy Butler photograph*)

West Chester, Pennsylvania, yard is the location of WCRR locomotive No. 6499 on October 4, 2004. The nearby area was once the location of the Pennsylvania Railroad turntable and freight yard. (*Jimmy Butler photograph*)

WCRR locomotive No. 1803 is at the West Chester, Pennsylvania, yard in 2004. The train consisting of the dining car (painted green), baggage car, and two coaches was heading back to West Chester returning from a special run to Glen Mills. (*Jimmy Butler photograph*)

In June 2005, WCRR locomotive No. 6499 has the former Pennsylvania Railroad baggage car behind it at the West Chester, Pennsylvania, yard. (*Jimmy Butler photograph*)

In June 2005, repainted WCRR locomotive No. 1803 is at the West Chester, Pennsylvania, station returning from Glen Mills with the engine cab facing forward for better visibility. (*Jimmy Butler photograph*)

This former Reading Railroad "Blueliner" is at the West Chester, Pennsylvania, yard in 2005 before it was repainted. The pantograph (still on top of the coach roof) was spring loaded and pushed up in a diamond-shaped configuration against the underside of the contact wire to draw the current needed to run the train. It was developed and patented by John Q. Brown of the Key System shops for their electric commuter operation that began operation on October 26, 1903 connecting San Francisco with the East Bay Area in California. (*Jimmy Butler photograph*)

In 2005, WCRR locomotive No. 1803 is at Glen Mills, Pennsylvania, station. The WCRR crosses nine roads (four in Chester County and five in Delaware County). There are nine bridges crossed by the railroad and one underpass at the Route 202 Bypass. (*Jimmy Butler photograph*)

WCRR locomotive No. 6499 is heading back to West Chester, Pennsylvania, on a December 2005 Christmas run. Winter snow creates a beautiful winter wonderland in the railroad's landscape and a great setting for the railroad's Christmas run. This is a railroad for all seasons as there is something different to be seen in each season. (*Jimmy Butler photograph*)

Left: December 2005 was the only year the WCRR had a Victorian Christmas with a Victorian Santa Claus alongside locomotive No. 6499 at Glen Mills, Pennsylvania, station. The railroad's Santa helps spread Christmas cheer throughout the holidays. Many families make the train an annual holiday event for a photo in the station with Santa and their family. (*Jimmy Butler photograph*)

Below: Passengers arrived at Glen Mills, Pennsylvania, station by WCRR train powered by locomotive No. 4230 for the 2007 Christmas season. Years ago, every day scenes like this at train stations were common place in the United States. (*Jimmy Butler photograph*)

WCRR locomotives Nos. 1803 and 4230 are at Clifton Aldan, Pennsylvania, station on a double header excursion to Philadelphia's 30th Street Station for the Springfield Historical Society of Delaware County, Pennsylvania in 2009. (*Skip Small collection*)

The Wiggins Christmas Tree Farm near Westtown, Pennsylvania, station is the location of a passenger train powered by WCRR locomotive No. 1803. (*Joseph Giacchino collection*)

The crossing gate is down for WCRR locomotive No. 1803 with its four-car train (one dining car, two passenger coaches, and one baggage car) to pass. The dining car can seat up to fifty passengers and has stationary tables for use during the trip. This train can be chartered, and the dining car can be added to enable patrons to have a memorable dining experience. (*J.M. Capra photograph*)

WCRR locomotive No. 1803 is at the West Chester, Pennsylvania, yard facing Glen Mills preparing to depart for 30th Street Station in Philadelphia in 2009. On the right in the distance, the WCRR volunteer conductor is walking back to reset the main line switch. (*Jimmy Butler photograph*)

WCRR volunteer Skip Small is by the crossing gate signal doing a safety check to make sure the WCRR locomotive No. 99 can safely cross the road. (*Skip Small collection*)

Christmas time is a great time to celebrate the holiday by having a train ride and enjoy the beautiful scenery such as in this view of WCRR locomotive No. 99 and passenger coaches at the Victorian Glen Mills, Pennsylvania, station. (*Joseph Giacchino collection*)

With the Christmas wreath on the front of WCRR locomotive No. 4230, everything is set for the next passenger run at the platform of the Market Street station in West Chester, Pennsylvania, which is the WCRR starting point. (*Skip Small collection*)

Magnificent-looking WCRR locomotive No. 4230 is powering a northbound train back to West Chester, Pennsylvania. On the right is the Wiggins Christmas Tree Farm, which is a short distance south of the Westtown station. (*Jimmy Butler photograph*)

Passing out of service Williamson School station in Middletown Township in Delaware County, Pennsylvania, WCRR locomotives Nos. 1803 and 4230 are double heading a rail enthusiast's excursion back to West Chester, Pennsylvania. That station opened in 1888 and closed after the last day of passenger train service to West Chester on September 18, 1986. (*Jimmy Butler photograph*)

On December 7, 2018, the WCRR station on Market Street in West Chester, Pennsylvania, is the northern terminus of the railroad and serves as a ticket office and gift shop. Outside the station is a 20-ton Plymouth locomotive greeting passengers and visitors waiting for the next train. The Plymouth locomotive was originally used for many years in the Glen Mills Quarry and was later donated. (*Ashton Krug photograph*)

WCRR locomotive No. 7706 is at the Westtown, Pennsylvania, station stop on October 23, 2018. The Westtown Station was built in 1857 and is one of the original stations on the railroad. It was built to handle passenger traffic for the Westtown School located about one mile east on Pennsylvania State Highway 926. A horse-drawn wagon provided transportation between the school and train station. Until 1883, the station was known as the Street Road Station, because Street Road is actually Pennsylvania State Highway 926. From March 1859 to the mid 1970s, Westtown Station was a U.S. Post Office, after which it was no longer a post office, and in 2020 is leased by artist Ken Kazanjian who has established the Westtown Station Gallery. (*Steve Rogers photograph*)

On October 23, 2018, WCRR locomotive No. 6499 is at picturesque Glen Mills, Pennsylvania. The first West Chester & Philadelphia Railroad train from Philadelphia arrived at Glen Mills on its way to West Chester on November 11, 1858. SEPTA discontinued passenger service on the Elwyn via Glen Mills to West Chester portion of the line with the last day of passenger train service on September 18, 1986. (*Steve Rogers photograph*)

WCRR locomotives Nos. 1803 and 4230 are at the West Chester, Pennsylvania, railroad yard on October 19, 2019 resting after a busy day's work. (*Kenneth C. Springirth photograph*)

At the West Chester, Pennsylvania, yard on October 20, 2019, the WCRR volunteer crew is using locomotive No. 4230 to make up the train on October 20, 2019. Yellow vests are worn by the WCRR volunteers to enhance safety on the railroad. (*Kenneth C. Springirth photograph*)

In the yellow raincoat, WCRR volunteer student brakeman Tyler Haney is preparing to step off to reset the switch off the main line at the West Chester, Pennsylvania, yard. The brakeman sets railroad track switches, couples, and uncouples cars to make up or break up trains. Signaling the engineer by hand signals, this job requires constant attention. On October 20, 2019, former Reading Railroad commuter coach No. 9107 and bay window caboose are being moved by locomotive No. 4230 to make up the train. (*Kenneth C. Springirth photograph*)

On October 20, 2019, WCRR locomotive No. 7706 is at Glen Mills, Pennsylvania, station waiting for departure time. The bricks on the pavement are all engraved with names of donors to the Thornbury Township Historical Society's renovation efforts to Glen Mills station. WCRR volunteer Peggy Morrison was heavily involved with the restoration of Glen Mills station and has handled ticket sales for the WCRR since it started on September 21, 1998. (*Derek Slifer photograph*)

On a gentle rainy October 20, 2019 at Glen Mills, Pennsylvania, station, smart-looking WCRR locomotive No. 4230 is waiting for passengers to board the train for the return trip to West Chester, Pennsylvania. (*Kenneth C. Springirth photograph*)

On October 20, 2019, the WCRR train powered by locomotive No. 4230 is back in the West Chester, Pennsylvania, yard after completing the scheduled trips. (*Kenneth C. Springirth photograph*)

On a sunny October 23, 2019, WCRR locomotive No. 1803 is ready to begin work for a busy excursion day at the West Chester, Pennsylvania, yard. (*Steve Rogers photograph*)

In 2020, the sun has set in this view looking towards West Chester, Pennsylvania, from the roof of WCRR locomotive No. 1803, the Budd built car (which serves as the WCRR office) is on the right, and a unique array of rolling stock and locomotives are now stored at the yard. (*Andrew L. Keates photograph*)

5
West Chester Railroad Stations

West Chester station, located on East Market Street between Matlack and Franklin Streets, opened in 1867. After a major January 2, 1885 fire, the two-story brick station was rebuilt. As ridership declined, condition of the station declined. With the closure of the news stand in 1962, only the ticket office remained. In June 1968, the main building was demolished and Penn Central Transportation Company built a passenger shelter at Nields Street, which was named West Chester University station. The West Chester station is the northern terminus of the WCRR.

Locksley station (actually a shelter) was established on May 24, 1890 based on a petition from Thornbury Township, Delaware County, residents. The name was chosen by a Pennsylvania Railroad division superintendant from a volume of poems by popular British poet Alfred Lord Tennyson (August 6, 1809 to October 6, 1892) who was also noted for famous English language phrases such as "The old order changeth, yielding place to new."

Glen Mills station, the southern terminus of the WCRR, is located at 130 Glen Mills Road in the community of Glen Mills in Concord Township, Delaware County, Pennsylvania, about 27 miles from Philadelphia. Dave Walter, historian of the West Chester Railroad Heritage Association (WCRHA) in an 1880 newspaper article found the current Glen Mills station (built during 1880–1881 with its elegant Queen Ann Gothic style) was designed by architect Joseph Wilson of the Wilson Brothers architectural firm who also designed the Morton station on the Pennsylvania Railroad Media–West Chester line. Restoration of the Glen Mills station began in 1985 by the Thornbury Historical Society with some financial assistance from the State of Pennsylvania. WCRHA volunteer Beth Anne Keates transcribed on May 24, 2020 the below Letter to Rich, a friend of the Thornbury Historical Society from an original cursive letter provided by John Anshant, secretary for the WCRHA.

Charles & Helen Hunt
Barnegat Light, NJ. 08006
March 28, 1987

Dear Rich,

Was delighted to receive your letter about the Glen Mills Station. Don't worry, you can hound me anytime. I was made P.R.R. agent at Glen Mills in 1951 and left in 1964 to take over the agency in Conshohocken, then Norristown, Rambo (King of Prussia), Earnest Yard near Norristown, then to Field Terminal Supervisor at Millville, New Jersey from where I retired in 1982 with job stress.

Altogether I had 38 years of service with the P.R.R., Penn Central and Conrail.

The railroad years ago wanted their agents to live in the stations. That's why there are quarters and homes on almost all of the stations in and around Philadelphia. Especially where they had block stations (Traffic Controllers) and sidings as Glen Mills had. We had block stations-we used the telegraph keys until 1950 in West Chester, Cheyney-part time, Glen Mills, Wawa then Philadelphia. Only on single track railroad did you have this, so you could run two trains against each other, one to the siding, the other thru, outbound trains from Philadelphia–Media took the siding; Philadelphia bound had the preference over the other traffic.

We brought up 4 children while at Glen Mills. It was a great life except for firing the huge coal furnace in the basement of the building, as you know it had 11' ceilings. I had to get up every morning and tend to the monster, at about 2 or 3 A.M. every night during the winter. Thanks to the railroad, they furnished the coal about 40 tons a year. Then you had to haul all of the ashes up and dump in the parking lot.

From what I can remember the station was built approximately 1860/1862 by then Baltimore, Wilmington and Washington Rail Road Company prior to becoming P.R.R. The line went from West Chester to Wawa, Chester, and Wilmington. The P.R.R. made the Philadelphia connection much later. Although P.R.R. freight still moved to Philadelphia and also to Chester via Thornton yard from Wawa to Lenni.

It had a slate roof, original, taken off in 1952/1953. They started in November 1952 and finished the day my oldest son was born February. I remember when hurricane Diane came thru. We were the only house that had water and heat. We had an old hand water pump in the back yard that I had to make sure the rail road plumbers kept in operation. Also we were the only house that was coal fired. The waiting room was filled with locals. We made the coffee (gas stove) and had a party. If my memory is correct, we were without electricity for 2–3 days.

The station also had a Ghost. An agent back around 1905 shot himself for stealing money. Many a night my wife would hear the steps up to the second floor make noises, as if someone was coming up the steps. I also heard it many times, but in time I guess he accepted us because of our kids. Also, while he was around the doors in the kitchen cupboards would be open, when we had closed them before going to bed. After a while it became a game.

I am enclosing some drawings of the station that may help you. I have old pictures we took years ago and when I get back to New Jersey, April 15th, I will look them up.

Don't worry, write when you want please. I will help in any way I can.

Sincerely,
Charles W. Hunt

P.S. Please keep me involved about your Progress. For your info in 1964, I asked the Real Estate agent of the P.R.R. to quote me a price to buy the station. I couldn't get up the $7500 to buy it. I should have but….

P.S. I just remembered you might check with Andy Wyeth as I know he sketched the station while I was there. Probably around 1960. Maybe his dad did also after all the station was 1 of a kind.

In 2001, WCRR owner Joseph Giacchino is the engineer on locomotive No. 3 at the Locksley, Pennsylvania station shelter, which is located about 0.125 mile west of Creek Road. (*Skip Small collection*)

WCRR locomotive No. 4213 is passing Locksley station shelter with an abundance of leaves on the ground in this autumn scene. (*Jimmy Butler photograph*)

On October 20, 2019, WCRR locomotive No. 4230 is at the Market Street platform awaiting departure time in West Chester, Pennsylvania. The building on the left is the WCRR gift shop plus ticket office put in place in April 2003. Largely funded through a Pennsylvania Department of Community & Economic Development grant, this is the first railroad passenger facility in West Chester since the June 1968 demolition of the Pennsylvania Railroad station. On the right is a stationary display of a Plymouth locomotive. (*Kenneth C. Springirth photograph*)

A 2005 snow-covered Christmas scene shows WCRR locomotive No. 6499 handling a passenger run at the Glen Mills, Pennsylvania, station. The scenic train ride plus looking at the magnificent Glen Mills station embellish the holiday season. (*Jimmy Butler photograph*)

The soothing sound of the clickedly clack of a train on the track attracts both adults and children as evidenced by this scene of WCRR locomotive No. 4230 in 2008 at Glen Mills, Pennsylvania, station. Often young children fall asleep on the return trip. (*Jimmy Butler photograph*)

The unique Glen Mills, Pennsylvania, station shines in the rain with WCRR locomotive No. 4230 ready to handle the trip back to West Chester on October 20, 2019. Characteristics of the Queen Ann Gothic-style architecture of this station include pointed arch windows, overhanging roof eves, upper decorated gable section above each dormer window (dormer is the window that projects out from the slanted roof), the deeply pitched roof (pitch of the roof is the roof slope), and the large wrap around porch. (*Kenneth C. Springirth photograph*)

6

West Chester Railroad Special Excursions

Since the first passenger train from Philadelphia reached West Chester on November 11, 1858, this line has offered scenic views and historic communities. From April to December, the railroad offers a variety of trips from their station located at 250 E. Market St., West Chester, PA 19382. The website is www.wcrailroad.com and their phone number during their operating season is (610) 430-2233. The railroad dining car (with large tables and seating for fifty-five guests) can be rented during a regularly scheduled train. An entire train can be rented for an event such as company events, conventions, family reunions, school trips, church trips, weddings, travel groups, rail hobbyists, community events, senior citizen outings, birthday parties for children and adults, anniversaries, graduation parties, and private events. Each of the railroad's passenger coaches seat seventy-five people. At the Glen Mills station, passengers can view the picturesque historic station before boarding the train back to West Chester.

A rare mileage trip was conducted on Sunday March 19, 2000 using two diesel locomotives from West Chester with a photo stop at Clifton Aldan and stop over at 30th Street Station and return to West Chester. Two special round trips were made between West Chester and 30th Street Station on Sunday March 14, 2004 in conjunction with the Chesapeake Railway Association.

On Saturday July 25, 2009, the WCRR in conjunction with the Springfield Historical Society and Morton Station Preservation Group operated a special trip from West Chester stopping at Morton on the SEPTA R3 Regional Rail line to 30th Street Station in Philadelphia. After it was turned at the Powellton coach yard, the train returned to Morton for a formal dedication of the saved outbound passenger train shelter. The train proceeded for a photo stop at Glen Mills, and then to West Chester with return back to Morton.

On May 17, 2001, a *Good Morning America* film crew visited the WCRR. On the left is the former Pennsylvania Railroad baggage car No. 9275, and on the right is the former Reading Railroad multiple unit commuter coach No. 9117 (with Fred W. Heilick, conductor, at the center of the doorway) at the station platform for the West Chester point of departure on Market Street.

On March 19, 1999, SEPTA's general manager plus staff, local government officials, and the WCRR volunteer crew celebrated the birth of the new railroad. Some of the people on the top row from left to right are Senator Thomas Thompson, Gene Cipriani-SEPTA legal department, and Joseph Kelly-SEPTA real estate department. Some of the people on the bottom row from left to right are Don Callander, Bob Bodan, Phil Klause, Andi DiNuzio, Ann Giacchino, Joseph Giacchino, Joe Lake, Bill Scott, Hank Habbersette, Ernie McNeely-West Chester Borough Council, Skip Small, Joe Parsons, and Glenn Schwass.

This former Reading Railroad and later SEPTA commuter coach is now celebrating a third career on the WCRR, and the special train event displayed a spirit of collaboration on March 19, 1999.

Trains have played a major role in the history, development, and preservation of the United States. Without the excellent railroad system that moved troops and supplies where needed, World War II might not have been won. You can have the best military; however, if personnel and supplies cannot be moved quickly, a battle can be lost. The WCRR made an excellent backdrop for a Chester County Republican Rally at the Market Street platform in West Chester, Pennsylvania, on November 18, 2000 with a banner tied to locomotive No. 99 arranged by Craig Lincoln Tucker of Chester County.

In this June 1998 crew photo of WCRR volunteers, Skip Small in a red shirt is standing (second person in from the end) on the front platform of WCRR locomotive No. 99 at Glen Mills, Pennsylvania. (*J. Nye photograph*)

A special WCRR train operated on March 19, 2000 from West Chester, Pennsylvania, via Lansdowne to 30th Street Station in Philadelphia. SEPTA helper locomotive No. 50 along with WCRR locomotive No. 1803 provided the motive power. In Philadelphia, the train was backed out and into Powelton Yard, and the locomotives were taken to Zoo Tower where they were turned. On the return as shown above, there was a photo stop at Lansdowne, Pennsylvania, on the Media-Elwyn rail line. (*Skip Small collection*)

Further up the ramp, this March 19, 2000 view shows the special WCRR excursion train returning from the trip to 30th Street Station in Philadelphia and a portion of the Lansdowne, Pennsylvania, station. Students of rail transportation had a chance to take pictures of the train at Lansdowne featuring SEPTA helper locomotive No. 50 along with WCRR locomotive No. 1803. Originally known as the Media-West Chester branch, the line was cut back to Elwyn with September 18, 1986 being the last day of operating passenger train service to West Chester. Since 1997, the heritage WCRR has operated between West Chester and Glen Mills. (*Skip Small collection*)

From left to right: WCRR locomotive No. 1803 and SEPTA helper locomotive No. 50 are at Clifton Heights, Pennsylvania on the special March 19, 2000 WCRR excursion from West Chester to Philadelphia. (*Skip Small collection*)

On July 4, 2000, antique automobiles from the Antique Car Club of America (which has many regions and chapters in Pennsylvania) are on display at the Glen Mills, Pennsylvania, station with WCRR locomotive No. 99 in the background. (*Skip Small collection*)

On October 27, 2009, SEPTA's Powelton yard (between the Powelton Village neighborhood of West Philadelphia and 30th Street Station in Philadelphia) is the location of WCRR locomotive No. 4230 with WCRR volunteer Shawn Dustman and a SEPTA conductor watching from the locomotive front platform on an excursion. (*Skip Small collection*)

The borough of Media (county seat of Delaware County, Pennsylvania) is the location of a special WCRR excursion train to Philadelphia with WCRR locomotives Nos. 4230 and 1803. (*Joseph Giacchino collection*)

On a June 4, 2011 excursion at Lansdale station, WCRR locomotive No. 1803 and SEPTA helper locomotive No. 70 (which provided power and safety to make this trip possible) are awaiting departure time. The station in the borough of Lansdale in Montgomery County, Pennsylvania, opened on February 7, 1903. This regional rail station was formerly an important transfer point between Reading Railroad electric commuter coaches and rail diesel cars built by the Budd Company. Bethlehem–Quakertown service ended July 1, 1981 and last day for Quakertown–Lansdale service was July 27, 1981. (*Joseph Giacchino collection*)

WCRR locomotive No. 1803 and SEPTA helper locomotive No. 70 are making a photo stop at Oreland station on the SEPTA Lansdale–Doylestown Regional Rail Line. Oreland is a village in Springfield Township, Montgomery County, Pennsylvania. The station was built by the Reading Railroad in 1931 to replace an earlier 1890 station. (*Joseph Giacchino collection*)

SEPTA double end Silverliner IV commuter car No. 402 (one of cars Nos. 270–303 and 400–416 built during 1974–1976 by General Electric Company) is passing locomotive No. 1803, leased from WCRR powering a wire train on SEPTA's Media-Elwyn Regional Rail Line. (*Joseph Giacchino collection*)

On June 4, 2011, at the SEPTA Powelton yard in Philadelphia, Pennsylvania, WCRR locomotive No. 1803 is passing SEPTA single end Silverliner IV commuter car No. 180 built by General Electric Company. (*Jimmy Butler Collection*)

Media–Elwyn SEPTA single-end Silverliner IV commuter car No. 316 (one of cars Nos. 101–188, 304–399, and 417–460 built by General Electric Company in 1974–1976) is passing a SEPTA wire train powered by leased WCRR locomotive No. 1803. SEPTA had leased No. 1803 for forty-one days from November to December 2003. (*Joseph Giacchino collection*)

SEPTA leased WCRR locomotive No. 1803 for several more years and used it as a work crew train working on the catenary all along the SEPTA electrified rail corridor. (*Joseph Giacchino collection*)

On June 4, 2011, WCRR locomotive No. 1803 and behind it No. 4230 are powering an excursion train for a photo stop at SEPTA's Powelton yard in Philadelphia for the return to West Chester. (*Jimmy Butler photograph*)

Double-headed WCRR locomotives Nos. 1803 and 4230 are using SEPTA's Media-West Chester line to get back to West Chester, Pennsylvania, on a June 4, 2011 excursion trip. (*Jimmy Butler photograph*)

Philadelphia's 30th Street Station is the location of the June 4, 2011 WCRR excursion train headed by locomotive No. 1803. It took an enormous amount of coordination and communication between SEPTA and the WCRR to make this trip possible. This was an all day trip with numerous photo stops. (*Jimmy Butler photograph*)

On June 4, 2011, WCRR locomotive No. 1803 is at SEPTA's Powelton yard (in Philadelphia) parallel to SEPTA car No. 9012 (one of seventeen Silverliner II cars Nos. 9001–9017 built for the Reading Railroad by the Budd Company in 1963). (*Jimmy Butler photograph*)

WCRR locomotive No. 1803 and SEPTA helper locomotive No. 50 are on a special excursion. (*Jimmy Butler photograph*)

The Lansdowne Avenue bridge provides the perfect spot to photograph a WCRR excursion train powered by locomotives Nos. 1803 and 99 at SEPTA's Lansdowne, Pennsylvania, station on the Media-Elwyn line in March 2000. This station, built in 1880, is located on the east side of Lansdowne Avenue one block south of Baltimore Avenue. (*Joseph Giacchino collection*)

A SEPTA helper locomotive and WCRR locomotive No. 1803 have paused at the Lansdowne, Pennsylvania, station in this photograph taken from the Lansdowne Avenue bridge in March 2000. (*Joseph Giacchino collection*)

On March 14, 2004, the WCRR West Chester–30th Street Station excursion featured an excellent photo stop at the Gladstone bridge over Darby Creek in Lansdowne, Pennsylvania, with locomotive No. 99 at one end and No. 1803 at the other end. (*Joseph Giacchino collection*)

WCRR locomotive No. 99 at one end and No. 1803 at the other end of the train are powering a special West Chester-Philadelphia 30th Street Station excursion near Highway 452 in Delaware County, Pennsylvania, on March 14, 2000. (*Joseph Giacchino collection*)

On June 4, 2011, WCRR locomotives Nos. 1803 and 4230 are passing the former Williamson School station on the Glen Mills to Elwyn portion of the Media-West Chester line that is currently out of service. This station is located on the grounds of the former Williamson Free School of Mechanical Trades, which opened October 31, 1891 by philanthropist Isaiah V. Williams to teach underprivileged young men a trade. On July 1, 2015, the school was renamed Williamson College of the Trades. (*Joseph Giacchino collection*)

WCRR locomotive No. 7706 is on a special excursion at the Amtrak Zoo Tower in West Philadelphia on September 27, 2016. (*Jimmy Butler photograph*)

WCRR locomotive No. 1803 and No. 4230 are at 30th Street Station in Philadelphia on June 4, 2011. This was part of the special train excursion to Lansdale, Pennsylvania. (*Joseph Giacchino collection*)

Wallingford station shelter located on Kershaw and Possum Hollow Roads in Nether Providence Township, Delaware County, Pennsylvania, is the location of a special WCRR excursion train, powered by locomotives Nos. 1803 and 99, which is returning from Philadelphia to West Chester on March 14, 2004. (*Joseph Giacchino collection*)

At the Moylan-Rose Valley station shelter on Manchester Road and Bancroft Road in Nether Providence Township, Delaware County, Pennsylvania, a photo stop is taking place of WCRR locomotives Nos. 1803 and 99 on the March 14, 2004 West Chester-Philadelphia round trip excursion. The brick station on the south side of the tracks was built in 1870.

Philadelphia's 30th Street Station is the location of WCRR locomotives 1803 and 4230 awaiting departure time for the return trip from Philadelphia to West Chester, Pennsylvania, on June 4, 2011. (*Joseph Giacchino collection*)

Passing Lenni station, WCRR locomotives Nos. 1803 and 4230 are crossing Lenni Road near Station Lane in Middletown Township, Delaware County, Pennsylvania, on the excursion train that is heading back to West Chester on June 4, 2011. This was on the portion of the West Chester line that SEPTA took out of service on September 19, 1986. (*Joseph Giacchino collection*)

In November 2005, Wayne Junction in Philadelphia, Pennsylvania, is the location of WCRR locomotive No. 1803, leased by SEPTA, pulling a work train. (*Jimmy Butler photograph*)

A special freight train photo charter Andrew L. Keates organized to raise funds for the WCRHA is crossing the road at Locksley station shelter in Delaware County, Pennsylvania, on May 6, 2018. On the right side of the photograph, a vintage Pennsylvania Railroad signal is still in place. (*Andrew L. Keates photograph*)

On September 27, 2016, WCRR locomotive No. 7706 is at SEPTA's Powelton Yard in Philadelphia ready to be turned at the wye at the Zoo Interlocking. SEPTA with its extensive commuter rail, subway/elevated, trolley, trackless trolley, and bus system has provided connectivity to the Greater Philadelphia region making it possible to reach employment, business, shopping, medical, educational, historical, and entertainment opportunities which have enhanced the region's quality of life. (*Derek Slifer photograph*)

SEPTA's 30th Street Station in Philadelphia is the location of SEPTA helper locomotive No. 70 type 2GS14B-DE and WCRR locomotive No. 7706 waiting to return to West Chester, Pennsylvania, on September 27, 2016. (*Derek Slifer photograph*)

Military re-enactors are lined up on the platform of WCRR locomotive No. 4213 at Glen Mills, Pennsylvania, station with the quarry master's house in the background on August 20, 2019. These re-enactors represented time periods from the Civil War to World War II. (*Beth Keates photograph*)

At this October 20, 2019 Trains & Soldiers plus author booksigning event, military re-enactors are standing on the side platform of WCRR locomotive No. 4230 at the Glen Mills, Pennsylvania, station, and the WCRR volunteer crew members (Skip Small, conductor; Frank Moscatelli, engineer; Ron Stevens; Dan Steil; Derek Slifer; Larry Jensen; Tyler Haney; Beth Keates; and John Anshantz) are standing alongside of the locomotive to symbolize the important ongoing working relationship railroads have had with the United States military. (*Kenneth C. Springirth photograph*)

7

Trolley Meets Train

The West Chester Street Railway Company (WCSR) began public operation of electric streetcar service on September 23, 1891 on two lines. The Market Street line operated on South New Street from Barnard Street to Market Street. It turned right on Market Street and operated on Market Street to the railroad station. The High Street line operated entirely on High Street between Virginia Avenue and Rosedale Avenue. Trolley car service from West Chester to Lenape opened on November 19, 1891 with a transfer point on the Wilmington & Northern Railroad at Lenape. Expansion of the line occurred with the opening revenue service on the line from West Chester to Downingtown on September 1, 1902. Construction of the line from Lenape to Kennett Square was completed in stages with revenue service reaching Unionville on July 23, 1904; Willowdale on September 20, 1904; Union Hill on October 19, 1904; and the Kennett Square borough line on November 12, 1904. The West Chester, Kennett & Wilmington Electric Railway (WCKW) on August 8, 1903 began trolley car service from Brandywine Springs, Delaware, to Kennett Square, Pennsylvania, and began laying track in Kennett Square to reach the WCSR at the Kennett Square borough line. That connection was completed and service from the South High Street depot to State and Union Streets in Kennett Square began on December 31, 1904. The WCKW also operated a trolley line from Kennett Square via Toughkenamon and Avondale to West Grove.

The Philadelphia & West Chester Traction Company (PWCT) began regular trolley car service from 63rd and Market streets in Philadelphia to Gay and Adams Streets in West Chester on January 4, 1899 and was extended to Gay and High Streets on June 1, 1899.

The WCSR began trolley car service from West Chester to Thorndale on August 26, 1906, and opened to Coatesville on November 2, 1906. Conestoga Traction Company (CTC) began through trolley car service from Lancaster to Coatesville on August 6, 1909.

Immediately PWCT (Philadelphia to West Chester), WCSR (West Chester to Coatesville) and CTC (Coatesville to Lancaster) coordinated schedules. Changing trolleys at West Chester involved walking one block between Gay and Market Streets. Changing trolleys at Coatesville involved walking via the Lincoln Highway over the tracks of the Philadelphia & Reading Railroad. Trolley riders could also change to the Pennsylvania Railroad commuter service at West Chester either via Media or via Paoli to reach center city Philadelphia. Passengers from Lancaster and Coatesville were more likely to use the Pennsylvania Railroad's mainline passenger service to reach Philadelphia.

In April 1923, the WCKW abandoned both the Brandywine Springs and West Grove lines. The WCSR took over the West Grove line in October 1923, immediately began to refurbish the line, and began through service from West Chester to West Grove on July11, 1924. As ridership declined with more people buying automobiles, the Kennett Square to West Grove trolley line was converted to Chester Valley Bus Lines route F on February 29, 1928. November 30, 1929 was the last day of WCSR trolley car operation.

On May 29, 1902, PWCT began trolley car service from 63rd and Market Streets in Philadelphia to Ardmore and was extended to the Ardmore business district on September 1, 1905. Philadelphia Rapid Transit Company (PRT) opened the Market Street Subway Elevated line from 69th Street Terminal in Upper Darby to 15th Street in Philadelphia. With the opening of 69th Street Terminal, PWCT began service from 69th Street Terminal to the Clifton–Aldan station of the Pennsylvania Railroad's Media West Chester line on April 30, 1907 and reached Sharon Hill on August 1, 1917. Service to Media opened on April 1, 1913.

PWCT became Philadelphia Suburban Transportation Company (PSTC) on April 13, 1936. Pennsylvania Highway Department plans for widening West Chester Pike resulted in converting the West Chester line to bus operation on June 4, 1954. Rush-hour trolley car service continued from West Gate Hills to 69th Street Terminal until it ended on August 23, 1958. The Ardmore trolley line was converted to bus operation on December 30, 1966. The Southeastern Pennsylvania Transportation Authority (SEPTA) took over PSTC on January 29, 1970. SEPTA equipped the Media and Sharon Hill lines with new light rail vehicles and upgraded the former Philadelphia & Western Railway into the Norristown High Speed line. The terminus of the Media line is within walking distance from the Media train station. The Sharon Hill line crosses under the Clifton-Aldan train station. Both stations are on SEPTA's Media-Elwyn line.

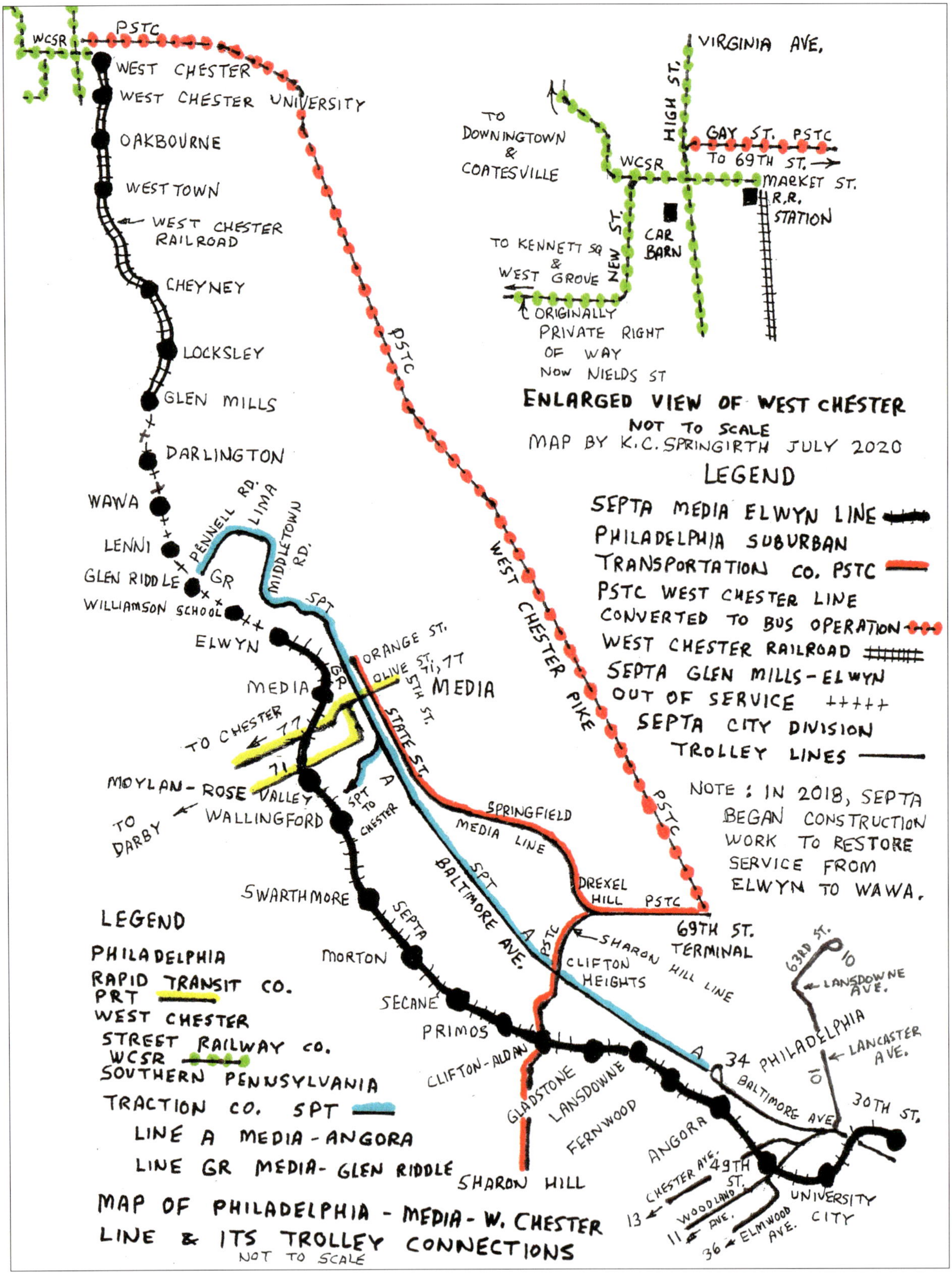

This map of the Philadelphia-Media-West Chester Line shows its trolley connections. Former trolley connections were: West Chester via PSTC to 69th Street; West Chester via WCSR to Downingtown, Coatesville, Kennett Square, and West Grove; Glen Riddle one-eighth of a mile up the hill to SPT via Lima to Media; and Moylan-Rose Valley via PRT Route 71 to Darby. In 2020, there are the following trolley connections: Media about a half mile to the Media trolley on State at Orange Street; Clifton-Aldan direct link to Sharon Hill-69th Street line; Angora to SEPTA Route 34 Baltimore Avenue; 49th Street to SEPTA Route 13 Chester Avenue; and 30th Street to SEPTA Routes 10, 11, 13, 34, and 36.

On May 20, 2007, this mural was photographed on West Chester Pike at Darby Road in Llanerch, Pennsylvania. On April 29, 1895, the "Battle of Llanerch Crossing" occurred when the Pennsylvania Railroad received a temporary injunction preventing the Delaware County Passenger Railroad from crossing its tracks at Llanerch for its new trolley line from 63rd Street to Castle Rock. On July 18, 1895 the injunction expired and the trolley crossing was completed. After a new company Philadelphia & West Chester Traction Company electrified the trolley line, over a period of time the struggle between train and trolley ended. (*Kenneth C. Springirth photograph*)

On West Chester Pike, Philadelphia Suburban Transportation Company (PSTC) car No. 84 was operating on West Chester Pike as an express to Llanerch, Pennsylvania. This was one of ten new one-man operated cars (Nos. 77–86) purchased by the Philadelphia & West Chester Traction Company (PWCT) from the J. G. Brill Company in 1932. The PWCT merged into PSTC on April 13, 1936. (*Jimmy/J. C. McHugh collection*)

Gay and High Streets in West Chester is the location of PSTC car No. 84. When the first PWCT trolley reached Gay and High Streets in downtown West Chester, Pennsylvania, on June 1, 1899, it provided the area with another way to get to Philadelphia. West Chester is the county seat of Chester County. (*Jimmy/J. C. McHugh collection*)

West Chester Pike was like a country road with only one automobile in sight when PSTC 80 series Brill car No. 82 made its trip via the side of the road private of way to West Chester. After the completion of the West Chester Street Railway (WCSR) from West Chester to Coatesville, Pennsylvania, on November 2, 1906, and Conestoga Traction Company (CTC) initiated trolley car service from Lancaster to Coatesville on August 6, 1909, it was now possible to travel between Lancaster and Philadelphia by trolley car. (*Jimmy/J. C. McHugh collection*)

Brilliner No. 3 is on the West Chester line before urban sprawl became commonplace along West Chester Pike. This was one of ten Brilliners built by J. G. Brill Company in 1941. (*Jimmy/J. C. McHugh collection*)

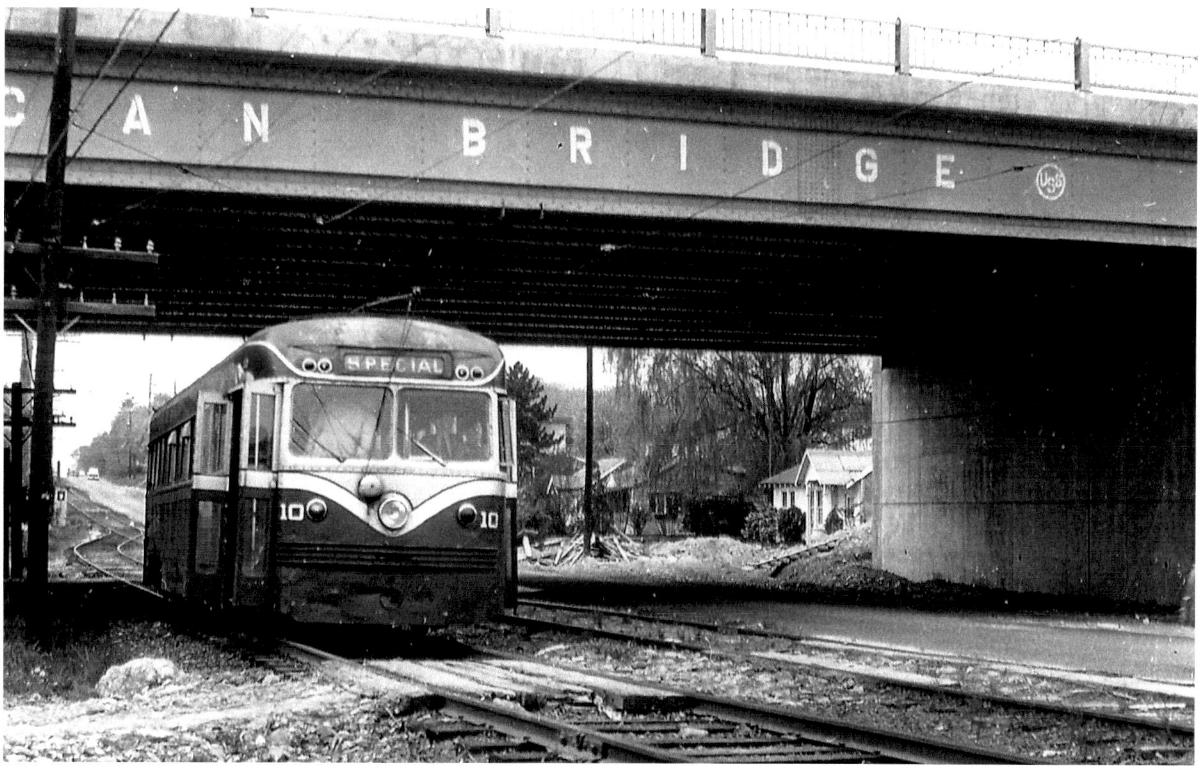

Brilliner No. 10 (with the destination sign showing "Special" indicating this is probably a rail enthusiast trip) is passing underneath the new Route 202 West Chester Bypass bridge that was built by the American Bridge Company founded in 1900 and headquartered in Coraopolis, a suburb of Pittsburgh. American Bridge Company built the Tappan Zee Bridge crossing the Hudson River at one of its widest points, Verrazzano Narrows Bridge connecting Staten Island and Brooklyn, and the San Francisco–Oakland Bay Bridge connecting San Francisco and Oakland. (*Jimmy/J. C. McHugh collection*)

This is a last run photo stop on a rail excursion on West Chester Pike at the Route 202 Bypass bridge built by American Bridge Company (ABC). McHugh Locomotive & Equipment had the contract to move bridge beams from the ABC Trenton, New Jersey, fabrication plant to East Coast job sites plus supplied the cranes for the West Chester job. On June 4, 1954, buses replaced the West Chester trolley cars. However, PSTC continued rush-hour trolley car service between 69th Street Terminal and Westgate Hills until that service ended on August 23, 1958. (*Jimmy/J. C. McHugh collection*)

At the western terminus of the Media line on State Street just west of Orange Street in Media (county seat of Delaware County, Pennsylvania) finds regular service PSTC Brilliner No. 1 and behind it a center door car that may have been on a rail excursion trip. The Media and Sharon Hill trolley lines have been retained and modernized by SEPTA. Less than a half mile walk south on Orange Street is the Media station of SEPTA's Media-Elwyn line. (*Jimmy/J. C. McHugh collection*)

An example of train and trolley connections offered by SEPTA's Media–Elwyn line is the Clifton–Aldan, Pennsylvania, station on March 14, 2004 where the commuter rail line crosses over SEPTA's Route 102 Sharon Hill trolley line. In this scene, a southbound Route 102 trolley car heading for Sharon Hill car is under the railroad bridge, a northbound Route 102 car is about to go under the railroad bridge, and WCRR excursion train with locomotive No. 99 has been carefully positioned making this picture possible. Vince Jakubowski, WCRR volunteer trip coordinator, arranged this photo stop. (*Joseph Giacchino collection*)

State Street just west of Orange Street in Media, Pennsylvania (terminus of the Media line), is the location of car No. 127. SEPTA acquired twenty-nine of these double-end light rail vehicles (modern term for trolley car) Nos. 100–128 built by Kawasaki Heavy Industries between 1980 and 1982. In 2020, these well designed cars ranging from thirty-eight to forty years old with their corrosion resistant construction have held up well. (*Kenneth C. Springirth photograph*)

At the 61st and Baltimore Avenue terminus of SEPTA subway surface Route 34 in the Angora neighborhood of Philadelphia, regular service Kawasaki Heavy Industries built car No. 9035 is awaiting departure time and partially hidden Presidents' Conference Committee car No. 2168 is on a rail enthusiast excursion on June 8, 2003. This location is close to the Angora station of SEPTA's Media-Elwyn Line. (*Kenneth C. Springirth photograph*)

On October 15, 2006, SEPTA Route 13 Kawasaki Heavy Industries car No. 9000 (built in 1980 and first of the 112 single-end light Rail vehicles (Nos. 9000-9111) delivered to the City Division) is on Chester Avenue at 49th Street in Philadelphia across the street from the 49th Street station of the Media-Elwyn Line. Route 13 operates to Yeadon with certain trips extended to Darby, Pennsylvania. However, the above car was signed 49th and Chester Avenue, because it was heading for the 49th and Woodland Avenue Depot. (*Kenneth C. Springirth photograph*)